Springer Series in Language and Communication 15

Editor: W. J. M. Levelt

Springer Series in Language and Communication

Editor: W. J. M. Levelt

Metalinguistic Awareness in Children

Theory, Research, and Implications

Edited by
W. E. Tunmer, C. Pratt, and M. L. Herriman

With Contributions by
J. Bowey R. Grieve M. Herriman M. Myhill
A. Nesdale C. Pratt W. Tunmer

With 6 Figures

Springer-Verlag
Berlin Heidelberg New York Tokyo 1984

Dr. William E. Tunmer
Dr. Michael L. Herriman

Department of Education, The University of Western Australia,
Nedlands, Western Australia 6009

Dr. Christopher Pratt

Department of Psychology, The University of Western Australia,
Nedlands, Western Australia 6009

Series Editor:

Professor Dr. Willem J. M. Levelt

Max-Planck-Institut für Psycholinguistik, Berg en Dalseweg 79
6522 BC Nijmegen, The Netherlands

ISBN-13: 978-3-642-69115-7 e-ISBN-13: 978-3-642-69113-3
DOI: 10.1007/978-3-642-69113-3

Library of Congress Cataloging in Publication Data. Main entry under title: Metalinguistic awareness in children. (Springer series in language and communication; v. 15) Includes bibliographical references and indexes. 1. Metalinguistic awareness in children. I. Tunmer, W. E. (William E.), 1947–. II. Pratt, C. (Christopher), 1950–. III. Herriman, M. L. (Michael L.). 1942–. IV. Bowey, J. (Judith). V. Series. P118.M45 1983 401'.9 83-19617

Softcover reprint of the hardcover 1st edition 1984

2153/3130-543210

To the memory of
David T. Hakes, 1934–1982

Preface

In the past fifteen years there has been a growing interest in the development of children's awareness of language as an object in itself -- a phenomenon now generally referred to as metalinguistic awareness. Until the publication of an earlier volume in the Springer Series in Language and Communication, *The Child's Conception of Language*, edited by A. Sinclair, R. J. Jarvella, and W. J. M. Levelt, there had been no systematic treatment of metalinguistic awareness. The major goal of that volume was to map out the field of study by describing the phenomenon of interest and defining major theoretical issues.

The aim of the present volume is to present an overview of metalinguistic awareness in children which reflects the current state of research and theory. The volume is divided into three major sections. The first considers various conceptual and methodological issues that have arisen from efforts to study metalinguistic awareness. It addresses such questions as what is metalinguistic awareness, when does it begin to emerge, and what tasks and procedures can be employed to assess its development in young children. The second section critically reviews the research that has been conducted into the four general types of metalinguistic awareness -- phonological, word, syntactic, and pragmatic awareness. In the final section the development of metalinguistic awareness is examined in relation to general cognitive development, reading acquisition, bilingualism, and early childhood education.

In addition to attracting the attention of those concerned with child language, this volume should be of interest to developmental psychologists, since the development of metalinguistic abilities appears to be related to the more general changes in cognitive capabilities that occur during middle childhood. The volume should also be of interest to educationists, since metalinguistic awareness is thought to play an important role in the acquisition of reading skills and in facilitating the difficult transition from preschool to formal schooling.

The chapters were written by members of an interdisciplinary research group, consisting of developmental psychologists and educationists, in the Departments of Education and Psychology at the University of Western Australia. The group's activities have been supported by both Departments and the University through the Tertiary Education Commission's General Development Grant, and by the Education Research and Development Committee. Earlier versions of some of the chapters appeared in a special edition of *Education Research and Perspectives* (1980, Vol.7, No.1). For the present volume they have been substantially revised and expanded. Other chapters, including the whole of the first section, are new. As a series of edited articles, the contributions reflect the views of those under whose name each appears. As well, each chapter has been written so that it may be read alone, and consequently some slight repetition of themes occurs across chapters. However, this has been avoided where possible by the use of cross referencing along with a common bibliography.

We hope that the chapters in this volume will provide clarity and direction to those interested in the topic, and that they will stimulate further study of the empirical and conceptual questions yet to be answered.

Nedlands, Western Australia
July, 1983

William Tunmer
Chris Pratt
Michael Herriman

Contents

List of Contributors

Bowey, Judith
Institute of Special Education, Victoria College,
Burwood, Victoria 3125

Grieve, Robert
Department of Psychology, The University of Western Australia,
Nedlands, Western Australia 6009

Herriman, Michael
Department of Education, The University of Western Australia,
Nedlands, Western Australia 6009

Myhill, Marion
Department of Education, The University of Western Australia,
Nedlands, Western Australia 6009

Nesdale, Andrew
Department of Psychology, The University of Western Australia,
Nedlands, Western Australia 6009

Pratt, Christopher
Department of Psychology, The University of Western Australia,
Nedlands, Western Australia 6009

Tunmer, William
Department of Education, The University of Western Australia,
Nedlands, Western Australia 6009

Part 1

General Considerations

1.1 The Development of Metalinguistic Awareness: An Introduction

Chris Pratt and Robert Grieve

The study of metalinguistic awareness in children has attracted considerable attention from developmental psychologists, developmental psycholinguists and educationists. In this chapter we will consider some of the reasons for recent interest in the topic and introduce some of the issues and problems that are associated with the study of metalinguistic awareness.

Metalinguistic awareness may be defined at the general level as the ability to think about and reflect upon the nature and functions of language. It is difficult to be more specific when defining the term, however, because the nature, functions and typical age of onset of metalinguistic awareness are still subject to much debate. An elementary reason for this debate is that psychology, in common with other disciplines, has not yet been able to provide a well articulated account of concepts such as awareness and consciousness which are involved in the study of metalinguistic awareness. At present it is not clear how different types, or different levels, of awareness might be distinguished, nor is it clear how these would relate to different degrees of consciousness. Furthermore, the attempt to relate such distinctions to the study of cognitive and linguistic processes generally is still in its initial stages, although recent work suggests a heuristic value for it.

In studying metalinguistic awareness in children, a large part of the task consists in trying to elucidate the phenomenon, both through conceptual analysis and empirical inquiry. Much of this book will be devoted to these issues and it will be shown that the conceptual problems in the area have not precluded important advances being made. Indeed the research into metalinguistic awareness has made an important contribution to our understanding of development during childhood. We will begin this chapter with a brief account of examples of the types of evidence that have been considered as indications of metalinguistic awareness in children. This will be done with reference to work conducted in three different research areas. It is by considering these different areas that some of the reasons for debate concerning the age of onset, types of awareness and phenomenon to be studied become evident.

The three areas that have led to specific interest in metalinguistic awareness in children are *early language development, learning to read,* and *cognitive development.* Since the second and third of these areas are discussed in later chapters they require only brief discussion here. The first area however, does not have a chapter devoted specifically to it and therefore is given more detailed consideration.

Metalinguistic Awareness and Early Language Development

Despite the great increase in research in children's language acquisition in the past twenty years, there is still much to be learned about the course of language development. Marshall and Morton [1978] comment that if we choose to represent normal language processes as being in a black box then the contents of the box are indeed "mysterious apparati". And further, when one considers the black box that is concerned with metalinguistic awareness, they point out that this box contains "even more mysterious apparati".

The issues concerned with metalinguistic awareness and its functions in early language development are complex and mysterious as Marshall and Morton point out. From both the theoretical and empirical standpoints there are difficult problems. With regard to the theory, there is continued debate about the role of conscious cognition in the acquisition of new skills. One view has it that it is at the time when one is acquiring a new skill that most conscious effort is directed towards it. We are as a result most aware of the processes involved at the time of acquiring a skill. According to this theoretical position, metalinguistic awareness would develop as the child is acquiring language. That is, an awareness of language would be intimately involved in its acquisition from the early stages.

However, there is an opposing view which holds that many skills are acquired without any conscious awareness of the processes involved. One commonly cited example is bike riding, where it is argued that beyond the rudimentary behaviours involved in sitting on the bike and pushing the pedals round, there is little awareness of what is involved in maintaining one's balance. In fact, a learner's attempts to try to think too deliberately about the processes involved in keeping his balance may often lead to disastrous consequences! (See Tunmer and Herriman, this volume, for further discussion of the relationship between awareness and skill acquisition.)

In the developmental context, Piaget [1928] argued that young children acquire many concepts spontaneously without having sufficient awareness of them to control or evaluate their use deliberately. Specifically, with respect to language and reasoning behaviour, Piaget argued that although young children may use connectives such as *because* and *therefore* correctly in their spontaneous speech, they have difficulty in using these words correctly when presented with a task that requires them to reflect upon their use. For example when asked to complete the sentence, "The man fell from his bicycle because...", children of 7 or 8 years of age would answer, "The man fell from his bicycle because he broke his leg" or "...because he broke his arm".

In developing Piaget's view Vygotsky [1962] draws a distinction between spontaneous or everyday concepts and skills, and scientific ones. Vygotsky argues that many everyday skills, including spoken language skills, develop spontaneously without conscious awareness of what is involved. Indeed, Vygotsky suggests that the awareness of the skill can only develop once the skill has been acquired. Hence, awareness of language does not develop until after language has been acquired and, Vygotsky argues, not until school age or later.

With respect to scientific concepts, Vygotsky argues that the awareness is highest when the skills involved are being acquired. Vygotsky uses the term *scientific concept* to encompass such skills as writing, reading, formal mathematics and laws of physics, skills which are typically formally taught to children in school.

In a situation where there are differing theoretical standpoints concerning the temporal relationship between the development of awareness and the acquisition of a new skill, it would be useful if there were some empirical evidence which would help to resolve the debate. In turning to consider such evidence from studies of child language, however, it becomes apparent that there are certain methodological problems which make it difficult to obtain a clear account of the development of metalinguistic awareness, particularly in younger children.

Researchers concerned with metalinguistic awareness and early language development have considered evidence from a number of sources. Clark [1978] has provided a review of much of this evidence which includes observations of children making corrections to language and making judgements about language. Since very young children cannot be expected to comment explicitly on the language they are just acquiring -- their limited language precluding them from doing so -- evidence for awareness of language in very young children often has to rely on inferences from what they may say and do. Consequently, much of the evidence relies on fortuitous observations of children's spontaneous comments rather than systematic inquiry using elicitation techniques. Such observations include speech repairs. Clark and

Andersen [1979] have argued that metalinguistic awareness plays an important role in enabling children to monitor the appropriateness of their utterances and in repairing their own speech. They suggest that if a child repairs his own speech spontaneously when it contains an error, then there must be some monitoring involved and that such monitoring implies a degree of awareness of the rules of language. Consequently in the following example from Snyder [1914] it may be inferred that the child, aged approximately two-and-a-half years, has some knowledge of word order: "Down sand beach I been - - I been down sand beach". But unless the child actually comments on the change it is difficult to tell whether repairs of this nature arise as a result of the child possessing *tacit* knowledge of instantiations of the rules of language rather than conscious knowledge of the rules themselves. That is, whether in Chomskian terms, it results from an underlying competence or tacit knowledge which enables the production of grammatical strings without necessarily implying an awareness of the rules involved.

The same problem concerning *tacit* knowledge of the rules of language also prevents clear conclusions being drawn from more formal evidence. In a study by Berko [1958] children displayed a knowledge of rules concerning the changes involved in pluralization and changing verbs into past tense. Thus with pluralization children were able to supply the final /-s/ to nonsense words (e.g., *wug* to *wugs*) which they had never heard before. But this work cannot be taken in itself as direct evidence of metalinguistic awareness in children because the task was not designed to elicit explicit discussion of the changes that were made to the nonsense words. Again, therefore, children's performances may reflect their tacit knowledge of the rules of language, rather than their conscious awareness of grammatical structure.

There are occasions when children do comment explicitly on speech corrections or other aspects of language however, as the following two examples from one child illustrate. In the first, Kate (2;6) was sitting on an adult's knee playing a game. The adult pointed to one of Kate's feet and asked, "What's that?" "A footsie", Kate replied. This was repeated with the other foot and the same answer was given -- "A footsie". The adult then pointed to both feet and asked "What are these?", and Kate replied "Two footsies -- no, two feetsies, *I mean*". The second example occurred some seven months later. Kate (3;1) was sitting at the table during a meal and asked for some cheese: "Can I have a bit of cheese, please? -- *Cheese please* -- that's a rhyme". In both these instances the child is making comments on aspects of the language being used. These comments would suggest some awareness of language in this child. However, the comments also underline problems in assessment. Consider what would have happened if in the above examples Kate had not made explicit comments in each case. It is certainly feasible that many children would not have. Kate generally has a good command of language and comes from a family environment in which linguisitic fluency is encouraged, whereas other children of a similar age may not even have such concepts as *mean* and *rhyme*.

Yet without the *I mean* in the first example, the evidence for some awareness becomes much less convincing, and similarly the lack of the reference to the rhyming words in the second example might leave it at best as an example of the type of children's word play similar to that documented by Weir [1962]. As the role of metalinguistic awareness in word play is not clear, this example would also be much less convincing.

Although there is anecdotal evidence from observations of children suggesting that some awareness of language develops early on, this has not been supported by systematic inquiry. In some cases systematic inquiry suggests that young children do not have the specific metalinguistic abilities being investigated. For example, in word tapping tasks where children are required to tap once for each word in two or three word strings, preschool aged children generally do not succeed in the tasks [see Bowey and Tunmer, this volume, for further details]. In other cases, however, where different metalinguistic abilities have been tested, problems have arisen which prevent firm conclusions being reached. For example, the results of a number of studies using judgement tasks [e.g., de Villiers and de Villiers, 1972; Gleitman et al., 1972; Scholl and Ryan, 1975] reveal the problems that may be encountered when children below 5 or 6 years of age are asked to judge the grammaticality of sentences. Despite ingenious attempts to convey to the child what is required in the task, younger children tend to focus on the content of the sentence rather than its grammatical structure. Thus, when asked to judge the grammaticality of the sentence "I am eating dinner" a 5-year-old in the Gleitman et al. study dismissed this sentence as unacceptable but gave as an explanation "I don't eat dinner any more". Further, many ungrammatical strings are thought to be acceptable by young children if they make sense to them insofar as the meaning conveyed by the string is consistent with their own experiences. However as discussed elsewhere [Pratt et al., in press; Tunmer and Grieve, this volume] a failure to elicit grammatical judgements from younger children does not necessarily imply that they are not capable of making such judgements. Although it may be the case that awareness of syntactic form has not developed, it is also possible that the techniques being used are not sufficiently sensitive to elicit grammatical judgements from younger children. Hence part of the challenge of future research will be to discover alternative ways of assessing metalinguistic awareness in children.

In concluding this section, it can be seen that there are differing theoretical stances on the relationship between awareness and the acquisition of language. Problems in deciding to what degree awareness is involved in examples such as language play, spontaneous repairs and sentence evaluation, are also evident. Partly as a result of such problems and partly because of different reasons for interest in metalinguistic awareness, much more attention has been paid to children aged 5 or 6 years and upwards.

Metalinguistic Awareness and Reading

The last two decades have seen dramatic changes in thinking about reading acquisition and reading instruction. With regard to the assessment of reading readiness for example, there has been a significant shift in emphasis from tasks designed to measure "perceptual readiness" (e.g., picture matching or identification tasks) to tasks designed to measure linguistic and metalinguistic skills. It has generally been found that performance on perceptual tasks does not provide a reliable indication of reading ability in children, and that in particular many children who are "perceptually ready" still encounter difficulties when learning to read. Because of these problems in assessing readiness, many researchers have turned to consider the links between spoken language skills and the acquisition of reading. Mattingly [1972] for example, has suggested that a lack of awareness of certain aspects of spoken language at the phonemic level (e.g., the ability to segment a word into its constituent sounds) results in children experiencing difficulties when they learn to read. That is, Mattingly proposes a link between awareness of certain aspects of spoken language and the child's progress when learning to read. His work, along with other research concerned with the links between reading and children's knowledge of spoken language [e.g., Calfee et al., 1973; Liberman, 1973], has provided impetus for further research into the development of metalinguistic awareness in children. Much of this research has been concerned with determining the precise nature of the relationship between reading and metalinguistic awareness. For although the work of Mattingly, Calfee and others suggests that at least some aspects of metalinguistic awareness are prerequisites for learning to read, alternative views have been expressed. Vygotsky [1962] and Donaldson [1978], for example, argue that the process of learning to read is in itself responsible for raising children's awareness of language.

Ehri [1979] presents another view, suggesting that there is an interaction involved such that a certain amount of metalinguistic awareness is required before the child learns to read and that learning to read will in turn increase the child's awareness of language. While such an interactionist view has certain appeal, it raises some challenging questions. Two important ones are; exactly which aspects of metalinguistic awareness are necessary prerequisites for learning to read, and what levels of awareness are required? If the aspects and required levels of awareness could be specified, then educators could provide children with more effective reading instruction.

Such considerations have led to much research into the development of metalinguistic awareness in children around the time they learn to read. In contrast with research into metalinguistic awareness and language development which has focussed on younger children, that concerned with reading has concentrated on children aged 5 years and upwards. This research has considered many different aspects of awareness including awareness of sounds, words, sentences, meaning and language terminology. For example, Bruce [1964] and Liberman et al. [1974] have investigated children's ability to segment words into their constituent phonemes while Downing and Oliver [1974], Bowey et al. [in press] and Tunmer et al. [in press] have examined the development of children's knowledge and awareness of words as units of language. Typically, such studies make use of experimental tasks designed to examine the specific aspects of awareness in question. Thus, for example, using a discrimination task, Bowey et al. required children to distinguish between one word and several words, and between one word and syllables. (See Bowey and Tunmer, this volume, for further details of tasks used to investigate word awareness.)

Research of this type has shown that there is an increase in various aspects of metalinguistic awareness around the age at which children learn to read. However, questions concerning the exact relationship between the development of awareness and learning to read, and the possible benefits of increasing children's awareness through training prior to reading instruction, have yet to be answered. One model which may prove to be of interest in pursuing such questions is outlined in the chapter on reading by Tunmer and Bowey [this volume], where specific aspects of awareness that are likely to play an important role at different stages of learning to read are identified. These range from a general awareness of language terminology and concepts of print during the early stages of learning to read, to the evaluation of the relationships that hold across propositions at a later stage of learning to read. This latter aspect, which is referred to as pragmatic awareness, is likely to play an important role in helping children to evaluate the extent to which they comprehend the material they read.

Further research will be necessary, however, to evaluate models of the relationship between metalinguistic awareness and reading. And although it may be possible to investigate the types of awareness involved, questions concerned with degrees or levels of awareness will continue to present the researcher with a difficult challenge.

Metalinguistic Awareness and Cognitive Development

Interest within the cognitive domain has been concerned with relationships between the development of metalinguistic awareness, other metacognitive abilities (e.g., metamemory, meta-attention) and children's cognitive development. Pratt and Grieve [this volume] consider in detail the hypothesis of Flavell [1978], Brown and DeLoache [1978] and others that the development of metacognition during childhood has a major impact on children's thinking. Metacognition refers to the ability to reflect upon and monitor the products of one's thought processes as a result of an increased awareness of the nature of cognitive functions. The development of such awareness is regarded as providing the child with much more control over his thinking. Donaldson [1978] has suggested that it is this development which enables children to succeed on many formal tasks used to assess their thinking, including such Piagetian tasks as conservation, classification and transitivity. Because there are important changes in children's performances during the period from 5 to 8 years of age, research into metalinguistic awareness from this perspective has also concentrated on children aged 5 years and upwards. Further, particular emphasis has been given to the self-monitoring skills which result from metacognitive development. In the area of language, this has led to research into children's ability to monitor their understanding of incoming information. Studies have included those by Markman [1977, 1979], Flavell [1981], and Flavell et al. [1981] which have been concerned with children's ability to evaluate the comprehensibility of instructions and stories. Much of this work rests on the assumption that an awareness of pragmatic aspects of language must develop in order to enable such evaluations to be made. However, there remain conceptual and empirical problems in ascertaining the exact relationship between knowledge or awareness of aspects of language and the influence of this awareness on performance. The development of awareness of language, for example, does not necessarily entail its application to monitoring the use of language in all contexts. For example, in a referential communication task, children may be aware that a good message should be unambiguous and should provide a listener with a clear description of the referent in question. However, there are many occasions when they do not apply this knowledge, and produce messages which remain ambiguous. Consequently, within the cognitive domain there remains a major question concerning the nature of the relationship between awareness and monitoring performance.

Summary

In this introductory chapter we have indicated the three major areas which have led to research interest in the development of metalinguistic awareness in children. In so doing, it has been noted that researchers have focussed attention on children of different ages, and have considered evidence of metalinguistic awareness in children obtained from various sources and in various ways.

With regard to age much of the research has focussed on the period of childhood from 5 to 8 years. This has mainly been a result of interest in relationships between reading and metalinguistic awareness, and cognitive development and metalinguistic awareness. However, the emphasis on this age group has probably also been influenced by the subjects themselves. Children of 5 years and upwards are easier to work with insofar as they are much more amenable to the procedures involved in experimental inquiry than are younger children. And it is with this type of inquiry that most researchers are comfortable. But techniques which will allow systematic study of metalinguistic awareness in younger children have yet to be fully developed. In many cases we may have to rely on good fortune in collecting the spontaneous comments which could indicate the development of some awareness. However, the lack of adequate techniques and systematic evidence at this stage cannot be taken as an indication of a lack of awareness in young children.

Further differences in the types of evidence considered by researchers also derive from the different interests they have in metalinguistic awareness. For example, many of those interested in early reading development have focussed on the child's awareness of phonemes and words, as it is these aspects of language which many believe play a major role in the early stages of learning to read. In contrast, researchers who have been interested in language acquisition have examined other aspects of awareness. In particular, interest in the development of child grammar has led to research into children's syntactic judgements. The aim here has been to determine whether young children possess a grammar different from adults. However, other problems concerning the basis of children's judgements in such tasks have prevented this aim from being pursued with very young children.

In the study of metalinguistic awareness in children then, there are conceptual and empirical issues which are relevant to all aspects of awareness. These include the conceptual problems of dealing with degrees or levels of awareness and consciousness, and methodological problems encountered in devising appropriate tasks and techniques. In addition to these general issues there are also problems that are specific to different areas of metalinguistic awareness. For example, in the study of metalinguistic awareness and cognitive development, the relationships between control and performance require careful examination.

In the following chapters in this book, many of the issues introduced briefly here are taken up in more detail. The following two chapters in this first section are specifically concerned with conceptual and methodological issues in the study of metalinguistic awareness.

The second section of the book contains four chapters devoted to consideration of four aspects of metalinguistic awareness. They are children's awareness of phonemes, words, sentence structure, and the interrelationships among propositions. Each of these chapters provides a review of the literature in the area and gives consideration to the theoretical and methodological issues of direct relevance to the aspects of metalinguistic awareness being considered.

The third section of the book contains four chapters that examine the role of metalinguistic awareness in different contexts. The first considers metalinguistic awareness and cognitive development, while the second considers metalinguistic awareness and the acquisition of reading. There follows a chapter on bilingualism and metalinguistic awareness, and a chapter on the educational implications of metalinguistic awareness.

Overall, the contents of the book reveal that despite the problems associated with the study of metalinguistic awareness in children much progress has been made. In recent years in particular, the interest and research which has focussed attention on the concept has helped to disentangle many of the issues involved. Whereas it would be remiss of us to fail to point out that our present knowledge of the topic is far from complete, the following chapters do indicate that significant developments have been made in recent years and that there is now a substantial amount of research in the area. Those concerned with the development and education of children will appreciate the importance of studying metalinguistic awareness. Finally, we feel that by drawing attention to some of the major issues and by providing a review of work in this area, this book should serve to stimulate further research into metalinguistic awareness in children, and to encourage further debate about the concepts involved in the study of metalinguistic awareness.

1.2 The Development of Metalinguistic Awareness: A Conceptual Overview

William E. Tunmer and Michael L. Herriman

This chapter examines the conceptual status of metalinguistic awareness and its relationship to other aspects of development in early childhood, namely, language acquisition, cognitive development and learning to read. As a first approximation, metalinguistic awareness may be defined as the ability to reflect upon and manipulate the structural features of spoken language, treating language itself as an object of thought, as opposed to simply using the language system to comprehend and produce sentences. To be metalinguistically aware is to begin to appreciate that the stream of speech, beginning with the acoustic signal and ending with the speaker's intended meaning, can be looked at with the mind's eye and taken apart.

Although metalinguistic awareness is related in meaning to the term *metalanguage*, it is important to distinguish between the two. While metalanguage refers to language used to describe language, and includes terms like *phoneme*, *word*, *phrase*, etc., metalinguistic awareness refers to awareness of the *instantiations* of these terms, but not to knowledge of the terms themselves. Thus, a metalinguistically aware child may perform well on a task involving the manipulation of phonemes without knowing what the term *phoneme* means.

The definition of metalinguistic awareness offered above gives rise to three possibilities regarding the epistemological status of metalinguistic awareness. To say that one is aware of linguistic entity X (e.g., a systematic phoneme) may imply that X *causes* the awareness, that X is merely the *intentional object* of the individual's thought, or that X is both cause *and* intentional object [Levelt et al., 1978]. In the sections that follow, we suggest that the latter position is probably correct, that linguistic entity X does, in fact, exist as part of the language processing system, and, in addition, can be the object of conscious reflection by means of invoking "control" processing.

Our procedure in this chapter will be to present a brief discussion of concepts that are closely related to and often confused with metalinguistic awareness. These preliminary conceptual distinctions will provide the background necessary for evaluating three views on the nature and development of metalinguistic abilities in children. Each view is distinguished at a general level by the age at which it is claimed that metalinguistic awareness typically appears. The first argues that metalinguistic awareness emerges at the *onset* of the language acquisition process, the second that it emerges *around* the time that formal schooling begins, and the third that it emerges *after* the child is introduced to formal schooling. In our examination of these positions we shall pay particular attention to the questions of what metalinguistic awareness is, what causes it, and what functions it serves.

The question of what metalinguistic awareness is, cannot of course, be settled by definition. The definition presented above should therefore be looked upon as a pretheoretical, working definition of the concept, the purpose of which is to provide us with some guidance in recognizing the relevant data. Ultimately, the question must be determined on the basis of empirical considerations. A major aim of this chapter, therefore, is to attempt to determine what the relevant data are and what they tell us about the hypothetical construct of metalinguistic awareness.

Preliminary Distinctions

In developing a conceptual framework for studying the emergence of metalinguistic awareness in children, it is essential first to distinguish it from four concepts that have evolved out of work in generative linguistics; *tacit knowledge, linguistic competence, linguistic intuitions,* and *explicit formulation. Tacit knowledge* refers to the unconscious knowledge that the speaker of a language has of the set of rules that determines the grammatical acceptability of the sentences of that language. A major task of the linguist is to develop a system of rules, or grammar, that represents the speaker's knowledge of his language, which linguists following Chomsky have called *linguistic competence.* This knowledge is unconscious in the sense that speakers are not aware of the rules they follow as they produce and comprehend utterances.

Not only are speakers unable to observe how the rules are utilized during ongoing speech production and comprehension, but they are typically unable to bring these rules into consciousness when reflecting back on what they said or heard. For example, a speaker may be able to say that the nonsense word *blench* could be an English word, whereas *bnench* could not, without knowing the phonological constraint in English which does not allow a word-initial stop consonant to be followed immediately by another stop consonant [Foss and

Hakes, 1978]. In making judgements about the language system, which are collectively referred to as *linguistic intuitions,* speakers generally are unable to provide *explicit formulations* of the rules underlying their judgements. Since linguists must infer the rules from the judgements that speakers make, as well as from the utterances they produce and comprehend, linguistic intuitions are an important part of the data that linguists use in constructing a theory of linguistic competence. In addition to judgements of permissible phonological sequences, linguistic intuitions include judgements of sentence acceptability, synonymy, and structural ambiguity.

Although linguistic intuitions involve metalinguistic abilities, they must not be equated with them. This is an especially important point to bear in mind in the evaluation of studies concerned with the development of metalinguistic abilities in children. It is entirely possible, for example, that children are able to perform metalinguistic operations without being able to provide explicit, adult-like judgements about language structure and function (i.e., linguistic intuitions). However, there have been successful attempts to elicit judgements characteristic of linguistic intuitions from young children when appropriate tasks were used [e.g., Hakes et al., 1980]. Other techniques and procedures for assessing children's ability to reflect upon language have also been developed and will be discussed in subsequent chapters.

In respect of awareness of the rules of language the child does not differ from the adult. Linguists accept that a speaker of the language has little or no knowledge of how his linguistic competence is represented in the cognitive system, or of the psychological processes involved in the utilization of this knowledge to produce and comprehend utterances and make judgements about them. As Levelt et al. [1978] point out, "neither the adult, nor the child, can become aware of the biological or mental machinery involved in language use" [p.6]. Rather, the speaker's awareness of his own mental processes is restricted to the *outcomes* or *results* of such processes [see also Nisbett and Decamp Wilson, 1977; Piaget, 1974a,b]. The speaker does, however, have *access* to the linguistic rules embodied in the subprocesses underlying sentence comprehension and production, where "access" is defined as the ability to form conscious judgements based on the operation of these internal mechanisms. Levelt et al. argue that how these judgements are made is an important question in its own right. It is this knowledge that has been referred to as *tacit* knowledge.

The suggestion that the speaker's awareness of language is limited to the *products* of the mental mechanisms involved in sentence comprehension and production leads naturally to the question; *what* structures are built up by the language user while interpreting or producing utterances; that is, what structures are "psychologically real"? An examination of the sound waves produced by a speaker reveals that they are nearly continuous. What we take to be clear segments of speech as denoted by the terms *phoneme, word, phrase*, etc., have no specifiable physical correlates. Consequently, there is no simple physical basis for isolating phonemes, words, or syntactic structures in speech. The available psychological evidence suggests, however, a model whereby the listener is able to *translate* the acoustic signal into a sequence of phonemes through a process referred to as speech perception. The process is a highly complex one, largely because there is no one-to-one correspondence between phonemes and segments of the acoustic signal. Information about phonemic content is transmitted in parallel, and the acoustic cues associated with particular phonemes vary considerably across different contexts [Liberman et al., 1967]. But despite the parallel transmission of phonemic content and the lack of invariance in acoustic cues, the speech perception mechanism, by a means not fully understood, produces as output a continuous series of phonemes, or "phonemic tape". This tape then serves as the input to another mechanism which (somehow) groups the phonemes and searches a mental lexicon to determine the meanings of the words in the utterance. The accurate identification of the individual phonemes is therefore essential, since the misclassification of a single phoneme can produce a word that is different in both meaning and form class (e.g., *r*an vs. *m*an).

Although the meaning of an utterance clearly depends on the meanings of the words it contains, it also depends on how the words are arranged, or parsed. Another processor, therefore, must take words retrieved from the lexicon and build a structural representation of them, from which the utterance's literal meaning is derived (for a review of the evidence supporting the existence of this intermediate stage in the processing of utterances see Foss and Hakes [1978]). Individual sentences, however, normally do not stand in isolation but are integrated into larger sets of propositions through the application of pragmatic and inferential rules. An intended, or integrated, meaning is therefore the final product of the sentence comprehension process.

In this brief sketch we have not intended to suggest that the processes carried out by the various postulated mechanisms do not overlap in time, or that there is no interaction among the mechanisms in the form of feedback of information from later mechanisms to earlier ones. Rather, we merely wish to argue in support of a model of comprehension which includes a number of *subprocesses*, each of which produces *outputs*. Although we have focused our attention on the *comprehension* of utterances, there is also evidence from research on language production, especially analyses of speech error data, which supports the psychological reality of underlying linguistic structures of the kind proposed in the model [e.g., Garrett, 1975]. These studies indicate that the structural parameters of language serve as important planning units in the production of speech.

As to the question of what structural features of language are psychologically real, the available evidence suggests that the products of the subprocesses involved in sentence comprehension include (at least) phonemes, words, structural representations of utterances, and sets of interrelated propositions. Metalinguistic awareness may thus be defined in information processing terms as the use of "control" processing (see below) to perform mental operations on the *products* of the mental mechanisms involved in sentence comprehension and production. This definition is consistent with the third possible interpretation of the epistemological status of metalinguistic awareness put forth by Levelt et al. [1978]; namely, that awareness of linguistic entity X implies that X is both the source of the awareness *and* the intentional object of the individual's thought. The structures that are built up during normal language processing are both psychologically real and subject to conscious reflection by means of control processing. Control processes, or metacognitive operations, will be discussed in a later section. They can be distinguished at a general level by the element of choice and relative slowness involved in the application of these operations.

Before concluding this section we shall briefly mention what metalinguistic awareness is *not*. Suppose a young child is observed uttering the sentence, "Mummy, bring me the cat". One might be tempted to argue that the child is "aware" of phonemes, since he used a phonemic difference to signal a difference in meaning [e.g., *c*at was selected, not *b*at]; that he is aware of words, since he used them to denote various actions, objects, and individuals; and that he is aware of syntactic structures, since he applied the *you* deletion rule in constructing the imperative form from the underlying structural representation of the sentence. However, this would be tantamount to saying that speaking itself is co-extensive with metalinguistic awareness, which would imply that the term is "totally vacuous" [Marshall and Morton, 1978].

An alternative view is that during normal language processing speakers use their (tacit) knowledge of the language to produce and comprehend utterances without being aware of the structures generated by the mental machinery involved in language use. That is, listeners are aware of hearing the sound of the speaker's voice and of understanding what was said, but are typically unaware of anything occurring in between. Similarly, speakers are aware of what they want to say, but not of the structures by which the content of their message is conveyed. In normal language processing, then, the language structures that are produced or derived between the acoustic signal and intended meaning are "transparent", to use Cazden's [1976] terminology. Accordingly, Cazden defines metalinguistic awareness as "the ability to make language forms opaque and attend to them in and for themselves" [p.603]. This metaphor for awareness of language is not new, however; Luria attributes to Vygotsky [1962] the "glass theory" whereby it is said that children see the world as through a glass (window). At a certain point in childhood they become aware of the glass itself. Cazden argues that this ability to make language opaque is a "special kind of language performance, one which makes special cognitive demands, and seems to be less easily and less universally acquired than the language performances of speaking and listening" [p.603].

A similar view is expressed by Mattingly [1972] in connection with learning to read: "Speaking and listening are primary linguistic activities; reading is a secondary and rather special sort of activity that relies critically upon the reader's awareness of these primary activities" [p.133]. Mattingly refers to this awareness of primary linguistic activity as "linguistic awareness", and argues that it varies considerably across speakers. He further observes that "this variation contrasts markedly with the relative consistency from person to person with which primary linguistic activity is performed" [p.140]. This raises the question of how and when metalinguistic awareness develops, a question that is addressed in the next section of the chapter, which presents a critique of three views on the development of metalinguistic awareness in children.

Three Views of the Development of Metalinguistic Awareness

View 1. Metalinguistic Awareness Develops Concomitantly with Language Acquisition

Arguments and evidence in support of this view are presented in Clark [1978], Clark and Andersen [1979], and Marshall and Morton [1978]. From the observation that children make spontaneous repairs to their speech from a very early age, Clark and Andersen conclude that children "are aware of language, its forms and functions, throughout the acquisition process" [p.11]. This conclusion, however, is predicated on the questionable assumption that "in order to make any repair, children *must* be able to reflect on their utterances so as to work out what has to be repaired on any one occasion" [Clark, 1978, p.23, emphasis added]. This assumption is certainly not true for adults, whose speech contains numerous false starts, self-interruptions, and changes of plan in mid-sentence. If adult speakers had to "reflect on their utterances" every time a repair was made, they would surely find it extremely difficult to concentrate on the content of what they were attempting to say. Nor does it seem likely that this assumption holds for children. Levelt et al. [1978] for example, speak of self-corrections as being "at the border of awareness" and suggest that "for the child speaker they may ... pass by on the stream of consciousness without leaving a trace" [p.2]. However, they note that "restarts can show that the child was aware that what he began or was about to say was inappropriate or incorrect" [p.2]. That is, young children may be aware that a mismatch has occurred between the meaning of the form produced and a stored representation of the one intended. But whether this leads to conscious reflection on the syntactic structure of the utterance in order to repair it, is another question, one to which we shall later return.

As a typical example of a speech repair, Clark [1978] cites the following from Scollon [1976, p.150], in which a girl aged 1;7 attempts to produce a recognizable form of the word *shoes:*

Brenda: (š)
(šI)
(š)
(šIš)
(šu)
(šu?)
(šuš)
Mother: Shoes!

In addition to phonological repairs Clark and Andersen [1979] describe three other kinds of spontaneous repairs made by young children -- morphological, lexical, and syntactic repairs. Morphological repairs include changes to the forms of pronouns (e.g., changing the case or person of a pronoun), and to the forms of verbs (e.g., adding the copula, correcting number agreement, or changing the tense or aspect inflection of a verb). Lexical repairs are those in which the child corrects an inappropriate word choice by substituting a second word for the first word chosen. Syntactic repairs are primarily made up of restarts, in which the child changes his choice of the subject noun phrase and backtracks to start again. An example of each of these three kinds of speech repairs is listed below:

1. (morphological) She want -- she wants to go to sleep.
2. (lexical) You have to squeak -- squeak -- scrape it.

3. (syntactic) The kitty cat is -- de-de spider's kissing the kitty cat's back.

After providing several examples of spontaneous repairs made by children as young as 2 years of age, Clark and Andersen briefly consider the role that these repairs might play in language acquisition itself. They suggest that for children in the early stages of acquiring language to realize that their elementary version of language is inadequate, they must become aware of when their language fails. When such failure occurs, the child must then reflect on his utterance in order to repair it. Clark and Andersen conclude that

> without the ability to monitor, check, and then repair one's utterances, it is unclear how children go about changing a rudimentary system into a more elaborate one. The mechanism of monitoring and checking seems to offer just the kind of mechanism that may be needed for the acquisition of such a complex skill as language. [pp.11-12]

A difficulty with this conclusion, however, is that there is now evidence that the ability to monitor, check, and repair one's utterances is certainly not a *necessary* condition for language development. Clinical cases have been reported of children who were unable to produce the sounds of human speech as a result of congenital defects in their speech production mechanisms. But despite this disability, the children showed essentially normal development in their ability to comprehend language [Lenneberg, 1962; MacNeilage et al., 1967]. It is therefore possible for a child to learn language without being able to produce speech, and hence without being able to monitor and repair his own utterances (see Foss and Hakes [1978] for a similar argument against imitation as a mechanism through which language is acquired).

Feedback mechanisms that monitor speech output may, however, be essential in learning to *produce* utterances. Marshall and Morton [1978] hypothesize that through the child's monitoring of his own speech, the perceptual system is able to "teach" the production system. In discussing the nature of the production system, they argue that the complexity of the processes involved is so great that devices for detecting malfunctions must be built into the system from the outset. Such devices must not only signal that something has gone wrong, but also where the malfunction has occurred and what kind of malfunction it is. Marshall and Morton maintain that metalinguistic awareness arises from these devices for detecting faults.

According to the theoretical model advanced by Marshall and Morton, normal language processing proceeds without awareness of the linguistic structures generated by the mental mechanisms involved in language production and comprehension. These mechanisms, the contents of which are "mysterious apparati", are to be distinguished from EMMA, an "even more mysterious apparatus", which monitors the results of the primary linguistic processes. EMMA is therefore defined as any device which monitors the results of the computations of the primary components of the performance system without itself being part of the primary performance machinery. Metalinguistic awareness in their view can be defined simply as Emma-functioning. That is, metalinguistic awareness results "from the operation of error-detecting mechanisms which have access to subparts of the output of primary production and comprehension systems" [p.237]. A consequence of this explanation of metalinguistic awareness is that there can be no *degrees* of awareness, since on any given occasion, EMMA either operates or does not.

In order for the comprehension system to teach the production system, the two systems must be linked together at some level. Marshall and Morton suggest that this occurs at the level of semantic representation (which, under their formulation, is essentially a combination of the literal and intended meanings discussed earlier). The comprehension system takes an acoustic signal and converts it into a semantic representation by a device called the compiler, whereas the production system formulates an expressible intention in a semantic representation, which is then converted into an acoustic signal by a mechanism called the expressor.

In discussing how EMMA operates in conjunction with these two primary systems to monitor and, if necessary, repair utterances, Marshall and Morton first consider the phonological repairs produced by Scollon's [1976] subject, Brenda (see earlier example), and those described by Clark and Andersen (e.g., "I mood -- I move it"). While agreeing that the latter kind of phonological repair may require some kind of internal monitor, Marshall and Morton maintain that "it is not necessary to think in terms of the child listening to her own output and correcting it on the basis of comparison with standard *adult* inputs" [p.234]. That is, such repairs do not imply an explicit awareness of phonological units on the part of the child.

Marshall and Morton then describe a feedback mechanism involving *semantic* monitoring which would appear to provide the means for correcting *all* the types of speech error mentioned by Clark and Andersen *without* explicit awareness of the products of the primary components of the performance system. If it is assumed that the compiler can take as input the child's own output, then a comparison can be made between the meaning of the form produced and a stored representation of the one intended, as indicated in Fig. 1. If a mismatch is detected, this would *automatically* result in the compiled version of the child's output being sent back to the expressor for re-encoding.

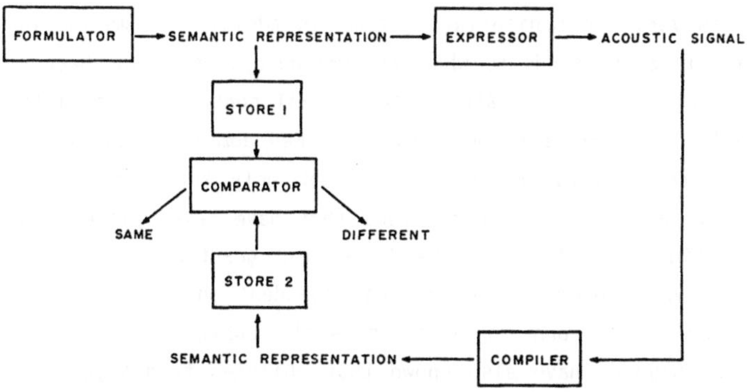

Fig.1. Feedback mechanism for correcting speech errors [after Marshall and Morton, 1978]

A somewhat more complex speech repair mechanism described by Marshall and Morton is one in which the compiler returns not only the compiled version of an incorrect string but also a copy of the particular rule (which could be phonological, syntactic, or semantic, as well as morphological) that was violated. On the assumption that the compiler first induces the rules by examining input strings (spoken in meaningful contexts), such "rule transmission" would enable the perceptual system to upgrade the production system and also provide the basis for an error correction process which would *not* require any conscious reflection on the structural parameters of language. Rather, this kind of monitoring system would only result in awareness (but not metalinguistic awareness; see below) when the speaker's output did not correspond to what he intended to say. When a mismatch did occur, it would trigger an automatic repair process in which the compiler sends back to the expressor for re-encoding the compiled output plus a copy of the rules or constraints that were violated. Although this would suggest that output monitoring is an automatic concomitant of normal language processing, it would not mean that the monitoring and repair mechanisms are not subject to the control of an "executive" routine. As Hakes [in press] points out, there are occasions, although rare, in which the speaker literally chooses his words carefully. Marshall and Morton would therefore appear to be correct in suggesting that the definition of metalinguistic awareness should perhaps be restricted to "EMMA plus consciousness", or, better yet, EMMA plus control processing.

It is possible to extend this version of the Marshall and Morton model to other language related activities involving feedback. Given that the purpose of speech acts is to convey intended meanings, it seems reasonable to suppose that speech repairs are motivated by the wish to make oneself understood. To achieve this goal, however, the monitoring system, in addition to monitoring output to ensure that it corresponds to what was *intended,* must also check to see whether the utterance is appropriate to the social context in which the speaker finds himself. By *context* we mean the presumed *common ground* shared by the speaker and interlocutor, which includes not only knowledge of the immediate situation but also of the interactions that have taken place up to that point, and anything else that can be taken for granted. Clark and Andersen [1979] cite several examples of repairs dependent on contexts, such as changing a pronoun to a noun to avoid possible confusion about the utterance's referent. Studies have also shown that children from 4 years of age are capable of adjusting their speech on the basis of the audience they are addressing or the role they have adopted [e.g., Shatz and Gelman, 1973; Andersen, 1977, cited in Clark, 1978]. It must be emphasized, however, that if social knowledge, in the form of pragmatic rules, is part of what the child knows implicitly about the language, then such (automatic) repairs and adjustments of speech output to make it congruent with the context of the situation are examples of the child *using* his language, not reflecting upon it. These cases are to be distinguished from problem-solving type communication tasks in which the child is required to consciously analyze, evaluate, and edit candidate messages [Flavell, 1977; Glucksberg et al., 1975]. Flavell [1977] suggests that these latter tasks require a fairly late-developing "metacommunicative" ability that involves "thinking about the message (metacommunication) rather than sending it (communication)" [p.178].

Another function of the feedback system is to make adjustments in the speech output when the listener has failed to understand what was said. Levelt et al. [1978] cite a study by Foppa [1978] in which a one-and-a-half year old child demonstrated the ability to cope with failures of speech acts. When the child spoke to its mother, the mother pretended not to understand by saying "mhm?" The child responded by restating the utterance with minor variations, and the entire procedure was repeated several more times. (See Käsermann [1980] for a detailed report on related studies using the same technique.) While these variations seem to indicate that "the child is aware that its utterance is in need of some kind of change or improvement" [Levelt et al., p.3], it is very difficult to imagine that a one-and-a-half year old child is able to consciously reflect upon the structure of its utterance in order to change it. The essential point here is that awareness of *failure* (resulting from the operation of EMMA) does not mean the same thing as awareness of *linguistic structure.*

A similar argument can account for the phenomenon of children correcting others. Using a criterion of contrastive stress to distinguish intentional from unintentional correction, Savic [1980] reports examples of 3-year-old twins correcting each other (see also examples cited in Clark [1978]). These findings, however, do not necessarily mean that these children are aware of grammatical structure, since they may have automatically "edited" each sentence while processing it, and subsequently detected a mismatch between the corrected, or compiled, version and the one that was actually spoken. Since the sentence actually spoken is momentarily stored in echoic memory, it is possible for children to compare it against the edited version and note any superficial differences without having to examine the grammatical structure of the sentence.

In addition to repairing and adjusting speech input and output, the feedback system is also involved in practising and playing with language. The distinguishing characteristic of verbal play is that it does not serve any immediate communicative function, since it typically occurs when the child is alone in his crib before falling asleep [Weir, 1962, 1966]. However, since verbal play often involves a recapitulation of some of the day's events, it is sometimes difficult to distinguish from speech repairs. An example of verbal play is when the child "practises" his pronunciation of particular sounds by producing sequences of rhyming patterns. If the rhymes were produced intentionally to create a particular phonological relationship, it would suggest that children as young as 2 years of age are able to reflect upon the phonological properties of spoken words. However, as Hakes [in press] points out, it may well be that in their bedtime monologues, children will occasionally produce sound sequences that do, in fact, rhyme, but which are not constructed deliberately: "It is difficult to discount the possibility that the rhyming that adults notice in young children's speech is more in the ears of the adults than in the mouths of the children" [p.42]. Just as children can use the words *bat* and *cat* to signal meaning differences without *knowing* that the two words differ only in their first phonological segment, so can they produce rhymes without knowing "phonologically" that *that* is what they are doing. Hakes argues that another reason for being cautious about interpreting young children's rhyming sentences as *deliberate* practice of phonological forms *per se* is that

> parrots engage in pronunciation practice remarkably like that observed in young children .... whatever one wants to say about parrots, it seems unreasonable to invest their practising with sounds with very much metalinguistic character. If so, is it more reasonable to do so for young children? [p.37]

In our discussion of the mechanisms involved in monitoring speech output, we distinguished between two types of awareness, awareness of failure and awareness of linguistic structure, and argued that only the latter type should be regarded as metalinguistic. Supporting this view is Piaget's [1974a,b] theory of awareness, which proposes that awareness develops initially from the first two conscious aspects of every intentional act, the aim and the result. In discussing Piaget's theory and its possible relevance to the development of language awareness in children, H. Sinclair [1978] states that "according to Piaget, in all intentional actions .... the acting subject is aware of at least two things: the goal he wants to reach and, subsequent to his action, the result he has obtained (success, partial success, or failure)" [p.195]. Since utterances are themselves intentional acts, the child, according to this view, should be aware of their purpose and results as soon as he begins producing them. Even the production of rudimentary forms of utterances implies that the child is consciously aware of the content or meaning of what he wishes to express, and of when his output fails to correspond to his intended meaning or fails to achieve the desired result in terms of the speech act he is attempting to perform (such as an assertion, question, request, command, etc.). Consistent with this view is the research (described earlier) by Foppa [1978] which indicates that children are aware of failures of speech acts from a very early age. According to Piaget's theory, however, awareness of failures to achieve the goals of intentional acts (*speech* acts, in this instance) is only the first step in "becoming aware". Conscious awareness (and control) of other aspects of the language system should therefore not appear until later in development.

Levelt et al. [1978] mention that awareness not only occurs at moments of failure to achieve the expected or desired result, but also during *skill acquisition*. The automatization of a complex skill such as riding a bicycle is preceded by a stage of conscious learning of the elementary skills required to perform this activity. However, as Levelt et al. point out, if it is true that a great deal of conscious attention is required to master the elementary skills involved in a complex action pattern, it would seem to lead to a paradox as far as metalinguistic awareness is concerned, since it would predict that young children in the process of acquiring language should be *more* aware of certain properties of language than older children, or even adults.

The apparent paradox can be explained in terms of the concepts of selective attention and short term memory. Consider what is involved in learning a new skill, such as riding a bicycle, which initially requires a considerable amount of concentration. Research has shown that short term (or working) memory (STM) is a limited capacity store, a "bottleneck" in the information-processing system. Learning a new skill places an enormous burden on STM, as the attentional mechanisms are not yet able to focus on that portion of the information flowing into the system that is relevant to the new task, and, consequently, cannot impose a structure on the incoming material prior to its placement in STM (a process referred to as "chunking"). As a result, STM is initially overloaded and the attentional mechanisms are under considerable strain. Although one may be generally aware of the strain, it is not possible to reflect consciously on the internal workings of the attentional mechanisms. When asked the question, "What did you think about when you learned to ride a bicycle?", most people would probably respond in very general terms, perhaps mentioning that their thoughts were trained on the *goal* (which would involve a conceptualization of successful bicycle riding) and the means for accomplishing it ("You hop on and start pedalling").

A similar argument may apply to language acquisition. The essential question is that, if young children are indeed paying a great deal of attention when placed in the role of hearer in a speech act, then to what are they directing their attention, if not to the structural features of the utterance? That is, how do they discover the linguistic rules which enable them to express their meanings? One approach to answering this question is to view the processes involved in acquiring language as reflecting aspects of general cognitive and social development. As Elliot [1981] puts it, "A child learning language is developing on all fronts, not just the linguistic one, and is trying to make sense of his social environment and the world of objects around him as well as his linguistic input" [p.37]. Recent research, summarized by Bruner [1978], indicates that children in the early stages of language acquisition are very adept at picking up clues to the intentions of adult speakers and using them to make sense of their utterances. This has led to the generally accepted view that children's early language comprehension performances are primarily interpretations of the utterance's context, rather than the utterances themselves, a view which renders "implausible an account which requires the child to reflect consciously on features of his language" [Elliot, 1981, p.38]. Rather, the child *consciously* attempts to ascertain the purpose of speech acts by interpreting the situation as a whole, while *unconsciously* attending to the relationships between the structural features of the utterances themselves and features of the situations in which they occur. That is, children construct linguistic rules through *unconscious* inferences from other speakers' utterances. As Chomsky [1965] has argued, "It seems plain that language acquisition is based on the child's discovery of what from a formal point of view is a deep and abstract theory -- a generative grammar of his language -- many of the concepts and principles of which are only remotely related to experience by long and intricate chains of unconscious quasi-inferential steps" [p.58].

The solution to the paradox described earlier thus lies in a distinction between two types of awareness, awareness of the goals (intended meanings) of speech acts and awareness of the structural features of utterances. The primary characteristic of intentional or goal-directed behaviour, which includes the production and comprehension of utterances, is that attention is trained on some desired end state (such as the communication of an intended meaning or the assignment of a coherent meaning to an incoming acoustic signal). When a child who is still in the process of acquiring language fails to comprehend an utterance addressed to him -- that is, when the *goal* of communication is not met -- the child relies on the context of the utterance to figure out what the speaker means, rather than reflecting upon the structure of the utterance. Similarly, since producing an utterance is an intentional act, the child is aware of the goal of the act, which is to convey the intended meaning. However, while young children are aware of the content of what they wish to convey in normal language production, they are not aware of the linguistic structures by which this content is conveyed. Only the latter type of awareness, which emerges later in development, is metalinguistic.

In concluding this section we wish to point out the way in which Marshall and Morton [1978] argue for the view that metalinguistic awareness develops along with language acquisition. In essence, they claim that it would be paradoxical for children to be "aware" of language without such awareness serving any useful function. "Given no useful role to perform, linguistic awareness becomes not so much the ghost within the machine as the ghost outside the machine" [p.227]. As a solution to this putative paradox, they attempt to "bring metalinguistic awareness back within the system" by arguing that it plays an essential role in language acquisition itself, in the form of feedback mechanisms that monitor speech output (see earlier discussion).

The argument, however, can be turned on its head. If children do indeed develop the ability to consciously reflect on and manipulate the structural features of language as a result of having acquired language, then how would this explain the great individual differences in metalinguistic abilities observed among children during middle childhood (the age range from 4 to 8 years), differences which contrast markedly with the relative consistency with which children of this age perform primary linguistic activities? Similarly, if children of this age are metalinguistically aware through having learned language (a process that is largely complete by 4 or 5 years of age), and if, as Marshall and Morton claim, EMMA operates in an all or nothing fashion so that there are no *degrees* of awareness, then why should so many children encounter difficulty in learning to read, a skill in which metalinguistic abilities are thought to play a major role? The answer, it seems, is that metalinguistic abilities do not develop concomitantly with language acquisition, but, instead, appear later in development.

View 2. Metalinguistic Awareness Develops in Middle Childhood and is Related to a More General Change in Information Processing Capabilities That Occurs During This Period

According to this view, metalinguistic awareness is a developmentally distinct kind of linguistic functioning that emerges during middle childhood. In support of this position is an accumulating body of research which indicates that middle childhood is the period during which the child is able to demonstrate a wide variety of linguistic skills which have in common the property of requiring the ability to reflect upon and manipulate the structural features of language [see the chapters in Section II of this volume for a review of this research]. Rather than simply using the language system to produce, comprehend, edit, or repair utterances, all of which are *automatic* processes, the language system itself is treated as an object of thought, with *control* processes being employed to perform mental operations on the products of the mental mechanisms involved in normal language processing. Metalinguistic operations therefore differ from normal language processes in the type of cognitive process exhibited. Automatic processes of the sort involved in normal sentence production and comprehension require little or no attention and are executed very rapidly, whereas control processes typically require a great deal of attention and involve an element of choice in whether or not the operations are performed [LaBerge and Samuels, 1974; Posner and Snyder, 1975; Schneider and Shiffrin, 1977; Shiffrin and Schneider, 1977].

For example, in comprehending an utterance, the listener is characteristically unaware that anything has intervened between his being aware of the speaker's voice and being aware of having understood the utterance, since processing normally proceeds in an automatic fashion. The listener does not notice such things as the individual phonemes and words comprising the utterance, the grouping relationships among the constituent words, or whether the utterance is structurally ambiguous or synonymous with another utterance, unless he deliberately thinks about it; that is, unless he invokes control processing to reflect upon the structural features of the utterance. Treating the language system as an object of thought, therefore, is not an automatic consequence of using the system as a vehicle for communication.

In addition to the type of processing involved, another distinguishing feature of metalinguistic operations is that their use tends to divorce language from its context. The metalinguistically aware child is able to abstract himself from the normal use of language and focus his attention upon the properties of language used to convey content rather than upon the content itself. He is able to analyse and manipulate aspects of language that previously were mastered only unconsciously by listening to and producing utterances in supportive contexts (see earlier discussion). As Donaldson [1978] argues, in the early stages of linguistic development, "The child's awareness of what he talks *about* -- the things out there to which the language refers -- normally takes precedence over his awareness of what he talks *with* ..." [pp.87-88]. That is, "before the child has developed a full awareness of language, language is embedded for him in the flow of events which accompany it" [p.88].

Evidence in support of this claim comes from several studies, some of which will be briefly mentioned here (see subsequent chapters for more detailed consideration of available research). Studies of the development of phonological awareness have shown that most 5-year-old children and many 6- and 7-year-olds are unable to segment spoken words into phonemes [Hakes et al., 1980; Liberman et al., 1974; Tunmer and Fletcher, 1981; Tunmer and Nesdale, 1982], even though four-week old infants can *discriminate* minor differences in speech sounds [Eimas et al., 1971]. Similarly, in a study of the development of young children's awareness of the word as a unit of spoken language, Tunmer et al. [in press] found that most children below 6 years of age were unable to segment meaningful phrases into their constituent words, responding, instead, on the basis of various acoustic factors. Consistent with this finding are several studies which indicate that children as old as 6 years encounter difficulty separating words from their referents [Markman, 1976; Osherson and Markman, 1975; Piaget, 1929; Vygotsky, 1962]. Rather than appreciating the arbitrary nature of the relationship between the meanings of words and their phonological realizations, young children tend to view words as inherent properties of objects, much like colour, shape and size.

A closely related finding has been reported by Hakes et al. [1980] who asked children ranging from 4 to 8 years of age to judge the acceptability of grammatical and ungrammatical sentences. While the older children tended to judge acceptability on the basis of the sentence's syntactic and semantic characteristics, the younger children were generally unable to dissociate the meaning of a sentence from its form. An examination of the reasons provided by these children for their judgements revealed that they tended to judge acceptability on the basis of the situations described by the sentences rather than on the forms of the sentences themselves. When the nature of the task is such that children are required to focus their attention on the structure of the sentence itself, such as in a task in which children are asked to "unscramble" sentences containing word-order violations that have rendered the original sentences meaningless, young children again tend to perform rather poorly [Pratt et al., in press].

The tendency of young children to be "empirically oriented" in their responses rather than focusing on aspects of the linguistic structure of the material has also been observed in studies concerned with the development of children's awareness of logical inconsistencies [Markman, 1979; Tunmer et al., in press]. Tunmer et al. asked 5-, 6-, and 7-year-old children to judge the acceptability of eight "stories" told by a toy puppet and found that the younger children experienced difficulty in detecting the logical inconsistencies contained within some of the stories. The justifications provided by these children indicated that they tended to question the empirical validity of individual sentences, rather than integrate the story as a whole and examine its overall logical structure.

These studies suggest that it is not until middle childhood that children develop the ability to deal with language in a "disembedded" manner, free from the supportive context of meaningful events. Metalinguistic abilities therefore appear to develop separately from and later than the abilities involved in sentence comprehension and production. The suggestion that metalinguistic awareness developmentally post-dates talking and understanding does not mean, however, that there are no parallels between the development of metalinguistic abilities and aspects of language acquisiton. As Karmiloff-Smith [1979a] points out, "A ... characteristic of language development after 5 appears to be the gradual passage from extralinguistic to intralinguistic reference, both in spontaneous utterances and, later, in metalinguistic awareness" [p.323]. A possibility that can be considered is that both late language development and the development of metalinguistic abilities are the reflection of an underlying change in cognitive capabilities that occurs during this period, or, as Karmiloff-Smith [1979b] describes it, a stepping-up to the "metaprocedural level".

Prior to age 5, during the preoperational stage of cognitive development, children apply general inductive strategies which lead to systematic errors in certain situations, such as in conservation tasks. This general tendency can also be observed in language development, where children initially resort to various strategies in comprehending complex linguistic structures. Examples of these strategies include the assumptions that the grammatical subject of a sentence corresponds to its logical subject, that the order in which events are referred to in sentences expressing temporal relations (e.g., before-after sentences) corresponds to the order in which the events occurred, that the implicit subject of a complement verb phrase is the noun phrase most closely preceding it, etc. (see Karmiloff-Smith [1979a] for a review of the research on developmental strategies). While such strategies generally lead to correct interpretations of syntactically simple sentences, their use with more complex structures often results in comprehension performance becoming *worse* with increasing age before it eventually improves, as Bever [1970] has demonstrated with reversible passives. The occurrence of a similar pattern of results in the development of conservation skills points to a possible relationship between cognitive and linguistic development during the preoperational period [Hakes et al., 1980].

This postulated developmental parallel between cognitive and linguistic development may also extend into the concrete operational period, as suggested by Karmiloff-Smith's [1979a] research on the acquisition of determiners by French-speaking children. She found that the children initially treated determiners as if they were unifunctional before gradually organizing them into a plurifunctional system. On the basis of her findings, Karmiloff-Smith concluded that "the gradual passage from juxtaposed, unifunctional homonyms, to plurifunctional systems of relevant options for modulating meaning, may be a general feature of development of the noun and verb phrase after 5 years" [p.322]. Moreover, these findings "illustrate the general trend for a representational tool which works well procedurally, to become subsequently an object of spontaneous cognitive attention per se" [Karmiloff-Smith, 1979b, p.97]. The increasing ability of children at the concrete operational level to deal systematically with relationships, enables them to come to grips with the intricacies of linguistic structures themselves, without having to rely on extralinguistic cues.

In addition to late language development being related to the major changes in cognitive development occurring during the concrete operational period, the same appears to be true for the development of metalinguistic awareness [Foss and Hakes, 1978; Hakes et al., 1980]. The abilities to *separate* a word from its referent, *dissociate* the meaning of a sentence from its form, and *abstract* oneself from the normal use of language in order to focus attention upon its structural features all sound very similar to what Piaget calls the ability to "decenter", or mentally stand back from a situation in order to think about the relationships involved. The essential feature of both metalinguistic awareness and concrete operational thought is the ability to *control* the course of one's own thought, which suggests that both may be the reflection of a more general change in underlying cognitive capabilities, the development of "metacognition" [see Flavell, 1977, 1978, 1981].

Flavell [1981] defines metacognition as "knowledge or cognition that takes as its object or regulates any aspect of any cognitive endeavour" [p.37]. It refers to the child's awareness of how he can control his own thought processes and involves the ability to perform mental operations on the products of other mental operations; that is, it involves "cognition about cognition". Recent research, reviewed by Flavell [1981], indicates that middle childhood is the period during which the child becomes increasingly aware of how he can control his own intellectual processes in a wide range of situations and tasks, including oral communication of information, oral persuasion, oral comprehension, reading acquisition, attention, memory, social cognition, and problem solving. Flavell [1977] says that

> As children develop, they generally become more capable of deliberately directing and controlling the deployment of their own attention. They become better able to focus their attention in an active, controlled fashion on just those external data which are relevant to their task objectives, while disregarding task-irrelevant data that are also present and equally perceptible .... such selectivity can occur at more than one stage of information processing .... a good attentional system will look where it wants to, discriminate and identify only the wanted portion of what it sees, and give full, cognitive attention (in perception, thought, or memory) only to the wanted portion of what it discriminates and identifies. [p.169]

Much the same is argued by Donaldson [1978]: "This question of control is at the heart of the capacity for disembedded thinking which...involves sticking to the problem and refusing to be diverted by knowledge, by beliefs or by perceptions which have nothing to do with it" [p.94].

Two empirical predictions can be derived from these considerations. First, from the suggestion that the broad variety of metalinguistic abilities that develop during middle childhood are the result of the emergence of a new kind of linguistic functioning (i.e., the ability to reflect upon and manipulate the structural features of language), it follows that children's performances on tasks requiring metalinguistic operations should be positively correlated with one another. Since different metalinguistic performances may each involve component skills that are unique to the particular task designed to measure it, we should not expect to observe the same *rate* of development across the different measures. However, a child whose performance is advanced on one should be advanced on all. The second prediction is an extension of the first. If metalinguistic performances reflect a common underlying change in cognitive capabilities -- that is, if each requires a component skill that is metacognitive in character -- then such performances should be correlated with performances on other kinds of *non*-metalinguistic tasks which also require metacognitive operations.

Evidence in support of these predictions can be found in studies by Hakes et al. [1980] and Tunmer and Fletcher [1981]. Hakes et al. measured 4- to 8-year-old children's ability to judge the acceptability of sentences, judge the synonymy of sentence pairs, segment words into their constituent phonemes, and solve standard Piagetian tasks. They found that the children's performances on the three tests of metalinguistic ability and one test of concrete operations were highly correlated. That the rather sharp increase in metalinguistic abilities reported during middle childhood coincides with the emergence of concrete operations indicates that the two changes may not be isolated events that just happen to occur during the same age period, but possibly are the manifestations of an underlying change in cognitive capabilities, the emergence of metacognitive control over the information processing system.

Further support for this claim is provided in a study by Tunmer and Fletcher on the relationship between conceptual tempo, phonological awareness, and word recognition in beginning readers. The tests used to measure conceptual tempo and phonological awareness, the Matching Familiar Figures Test and the phoneme tapping task, each involved a number of component skills, one of which was common to both, namely cognitive control. The two tests were administered along with tests of reading ability and intelligence, to a sample of first graders. It was found that the children's scores on the two tests were significantly correlated, even when the effects of intelligence were held constant. Tunmer and Fletcher concluded that, since the general characteristics of the two tests were completely dissimilar, with each involving a different perceptual modality, a significant correlation between the two would not have been expected unless the two tests shared a common underlying feature in cognitive processing.

Since both of the studies we have described were correlational ones, it is possible, of course, that the obtained empirical relationships were due to some factor other than controlled cognitive processing. Some researchers find it tempting to link the striking changes in metalinguistic and cognitive abilities observed during middle childhood to the effects of formal schooling [e.g., Donaldson, 1978], particularly in view of the fact that such changes occur around the age at which formal schooling begins. The arguments relating to this position are presented in the next section of the chapter.

View 3. Metalinguistic Awareness Develops After the Child Begins Formal Schooling and is Largely the Result of Learning to Read

One of the main proponents of this view is Donaldson [1978]. She argues that introduction to formal schooling, especially learning to read, brings about an increase in metalinguistic abilities which, in turn, enables the child to exercise greater control over his thought processes so that he can use cognitive skills in a wider range of situations. Donaldson therefore lays great stress on the early acquisition of reading skills as a means of facilitating the difficult transition from "embedded thought" to "disembedded thought": "...I believe that the *early* mastery of reading is even more important than it is commonly taken to be" [p.96, emphasis added]. To accelerate the growth of their cognitive capabilities, children should be introduced to reading instruction as soon as possible after beginning formal schooling.

This thesis raises conceptual and empirical problems. While research does indeed show that there are positive correlations between measures of metalinguistic awareness and reading achievement, this does not answer the question of the direction of cause and effect. On conceptual grounds it would appear that the development of metalinguistic awareness is a *prerequisite* for being able to learn to read. The child's fundamental task in learning to read is to discover how to map the printed text onto his existing language, a task which requires the ability to deal explicitly with the structural features of the spoken language [Mattingly, 1972; Tunmer and Bowey, this volume]. Without the metalinguistic ability to reflect upon language the child would not be able to discover the properties of spoken language that are central to the correspondences between its written and spoken forms. This would suggest that metalinguistic awareness is a necessary, but not sufficient, condition for learning to read. We should therefore find children who are metalinguistically aware, but cannot read; but not find children who are not metalinguistically aware but are able to read. Supporting this claim are the results of the study cited earlier by Tunmer and Fletcher [1981]. They found a relationship between phonological awareness and reading ability among beginning readers, but one that was nonlinear. There were several children who were phonologically aware but did not perform well on a synthetic word test (presumably because they had not yet mastered the grapheme-phoneme correspondence rules), but no child who performed poorly on the phonological awareness task but well on the reading task. Similarly, the findings of Ehri [1979] and others indicate that many preschool children who have not been exposed to reading instruction and who cannot read, perform well on tests of metalinguistic awareness that correlate highly with reading achievement. And several studies report that the scores of kindergarten children on a variety of tests of metalinguistic ability are significant predictors of reading achievement in first grade [see Ehri, 1979, for a review of available research].

Donaldson's analysis does not explain why so many school children (10 to 20 per cent, according to Downing [1973]) who have no apparent visual, hearing, or mental deficit encounter unusual difficulty of one kind or another in learning to read, regardless of the method of teaching reading used. Nor does it account for the fact that children who are not taught to read (i.e., those living in nonliterate societies) follow much the same pattern of cognitive development, as outlined by Piaget, as do children from literate societies [see Cole and Scribner, 1974]. Learning to read, therefore, seems not to be a *necessary* condition for the development of either disembedded thinking or metalinguistic awareness.

The alternative view to Donaldson's thesis, namely, that metalinguistic awareness is a prerequisite for learning to read, is not inconsistent with the possibility that reading *instruction* increases metalinguistic awareness [Byrne, 1981]. Most types of reading instruction include some form of language awareness training, which would explain the sharp increases in metalinguistic abilities often observed among beginning readers. As Donaldson [1978] herself argues, "the *process* of becoming literate can have marked...effects on the growth of mind" [p.97]. However, it may also be true that children must reach a certain "threshold" level of cognitive development, of the sort described in the preceding section, before they can benefit from reading instruction; that is, before they can perform the metalinguistic operations necessary to acquire the component skills of reading. If some children develop metacognitive control *before* they begin formal schooling while others develop this ability afterwards, it would explain why so many children encounter difficulty in learning to read, and why some children spontaneously develop the ability to treat language as an object of thought prior to formal schooling. Although Donaldson also maintains that there is a close relationship between the development of metalinguistic awareness, disembedded thinking (metacognitive control), and learning to read, she argues for a somewhat different causal sequence; namely, that learning to read gives rise to metalinguistic awareness which in turn results in disembedded thinking.

Rather than attempting to teach the child to read earlier, a better strategy might be first to prepare the child for this difficult undertaking by training him on those metalinguistic skills which are essential for learning to read [Weaver and Shonkoff, 1978]. Donaldson makes a similar point in suggesting that "the preparation for reading should include, as a most important component, attempts to make children more aware of the spoken tongue" [p.97]. A few training studies on different aspects of metalinguistic awareness have been conducted and indicate that the development of metalinguistic abilities can be facilitated by training, and that such training transfers positively to reading achievement (see subsequent chapters in this volume for further discussion). If metalinguistic awareness were merely a consequence of learning to read, training to increase metalinguistic awareness should have no effect on the course of reading acquisition.

Concluding Remarks

In this chapter we have attempted to provide a general theoretical framework for integrating the available research on the development of metalinguistic awareness in children. We began by conducting a conceptual analysis of the notion of metalinguistic awareness which resulted in its being defined as the use of control processing to perform mental operations on the products of the mental mechanisms involved in sentence comprehension and production. Three views on the nature and development of metalinguistic awareness in children were then considered. The first holds that the development of metalinguistic awareness is part and parcel of the language acquisition process, while the other two see it as separate from the acquisition of language itself. The third view holds that the awareness of language is largely an effect of learning to read. We suggest that on the basis of the conceptual models and empirical research currently available, the most supportable view is the second, namely, that metalinguistic awareness is a developmentally distinct kind of linguistic functioning that emerges during middle childhood, and one that is related to a more general change in information processing capabilities that occurs during this period.

1.3 The Development of Metalinguistic Awareness: A Methodological Overview

Andrew R. Nesdale and William E. Tunmer

Although the study of the development of children's metalinguistic awareness is still in its infancy, a surprising amount of research has already been accomplished [see reviews in this volume by Nesdale et al.; Bowey and Tunmer; Tunmer and Grieve; Pratt and Nesdale]. Nevertheless, as might be anticipated given the newness of the field and the complex and multifaceted nature of the construct, refined and generally accepted methods of assessing metalinguistic awareness are still in the process of being developed. To date, the methods which have been used by researchers have tended to vary quite markedly between studies and little attention has been given to the stability of children's responses on the same task and procedure over a short term. In addition, the research conducted has tended to be cross-sectional. Together, these factors have contributed to the area being notably devoid of reliability and longitudinal data. And, they have reduced the possibility of close comparisons being made between studies utilizing different methods in order to determine which elements contributed to different levels and/or types of performance. In most areas of research on children's metalinguistic awareness at the present time, it is unclear whether conflicting estimates revealed by different studies are due to sampling, task, procedural or measurement differences. At best, the available research allows for only imprecise estimates of, for example, the age at which a particular facet of metalinguistic awareness emerges and the period of its development.

There are several reasons why it is important to provide precise estimates of the course of development of metalinguistic awareness. First, one aspect upon which the current views of metalinguistic awareness can be differentiated concerns the age of emergence of this ability [see Tunmer and Herriman, this volume]. Clearly, establishing the age of emergence of this ability is dependent upon the development of valid methods of diagnosis. Secondly, given that there are good grounds for considering that other abilities, such as reading, are dependent upon the development of metalinguistic awareness [see Nesdale et al., this volume; Tunmer and Bowey, this volume], the possibility of devising programs to enhance metalinguistic awareness becomes an important consideration. Again, the evaluation of training research is dependent on accurate measures of pre-test and post-test performance.

While it is not possible to provide an algorithm for a precise measure of metalinguistic awareness at present, the picture is not quite as bleak as the initial paragraph might suggest. The research to date has at least led to the recognition of, and consideration being given to, a number of important and thorny methodological issues. While some of these issues appear to be unique to the study of metalinguistic awareness, others are similar to those faced by any researchers working with children, especially when the focus is on the child's linguistic, cognitive or metacognitive abilities [e.g., Brown, 1980; Cavanaugh and Perlmutter, 1982]. An awareness of these problems together with systematic attempts to overcome them should increase the likelihood of more reliable and valid data being obtained and hence of a more accurate picture of the sequence of development of metalinguistic awareness being revealed.

The present chapter consists of three main sections. In the first two sections, we consider the methodological implications of two important conceptual issues concerning metalinguistic awareness. These include the measurement implications of, firstly, the different theoretical positions regarding the emergence of metalinguistic awareness and, secondly, the multifaceted nature of metalinguistic awareness. This discussion serves to raise the important distinction between task and procedure. In the third section, the discussion focuses on several of the general methodological issues relating to research on metalinguistic awareness including the instructions given, the use of practice trials and feedback, and the responses required to demonstrate competence.

Emergence of Metalinguistic Awareness

Although it is generally agreed that metalinguistic awareness refers to the ability to reflect upon and manipulate the structural features of spoken language [Tunmer and Herriman, this volume], there is considerably less agreement concerning *how* and *when* metalinguistic awareness emerges and *what* behaviors may be taken as an index of it. Tunmer and Herriman have examined three views of metalinguistic awareness which differ markedly on one or more of these three factors. In brief, according to one conceptualization, metalinguistic awareness is an integral part of the process of language acquisition and is hence acquired early in life. Spontaneous speech repairs and language play are cited as the primary evidence in support of this position. In contrast, according to a second view, the development of metalinguistic awareness occurs during middle childhood, the period from approximately 4 to 8 years of age. It is considered to reflect a new kind of linguistic functioning which is influenced greatly by the cognitive control processes which begin to emerge during this period. The operation of metalinguistic awareness is again considered to be revealed in a variety of behaviors. However, compared with spontaneous speech repairs and language play, these are considered to be characterized by controlled and intentional processing (e.g., being able to segment a sentence into words, and words into phonemes) rather than by automatic processing. According to the third position, the development of metalinguistic awareness is also considered to post-date language acquisition but, in contrast to the second position, it is thought to be largely the result of exposure to formal schooling, especially learning to read. As with the second view, metalinguistic awareness is considered to be evidenced in a variety of controlled responses.

These different conceptualizations of metalinguistic awareness have important methodological implications. One such implication concerns the *age* at which children begin to demonstrate metalinguistic awareness and the *age range* over which this ability develops. According to the first approach noted previously, the researcher would expect to find evidence of metalinguistic awareness when the child commences speaking at approximately 18 months of age whereas, according to the third position, researchers would expect to observe the emergence of metalinguistic awareness just after the commencement of formal education. The implication which follows is that researchers would need to develop techniques which are appropriate to the age of the children being tested. Moreover, to assess the different conceptual viewpoints, techniques would need to be developed which span the age range from approximately 18 months to 8 or more years of age.

In addition, the three approaches have different implications concerning the *range of tasks* upon which the child would be required to demonstrate some level of competence in order to be credited with an amount of metalinguistic awareness, and the *procedures* which are used to assess the child's competence. It will be noted here that a distinction is being drawn between *what* the child is required to do or demonstrate (i.e., the task) and *how* this requirement is to be assessed (i.e., the procedure). Both demand special attention from the researcher, although the latter presents greater difficulties. Thus, in terms of the range of tasks, the second and third approaches would, for example, assess the child's phonological and word awareness but would be less concerned with examining the child's tendency to repair his speech spontaneously. While the task analysis would create no special problems for the researcher, difficulties do arise when the researcher attempts to devise procedures appropriate for assessing the competence on the different tasks of the children in the different age groups, especially the younger children. Clearly, in the interests of obtaining stable and valid estimates of a child's ability, the researcher would prefer to test the child in a controlled setting using standard procedures (i.e., fixed instructions, stimuli and response measures), rather than rely on analyzing and interpreting data obtained via an uncontrolled observation technique (e.g., spontaneous speech production recordings or an unsystematic sample of anecdotes). As Gleitman et al. [1972] have pointed out, "the spontaneous speech of children provides a limited data source for the study of their linguistic knowledge, in practice as well as in theory" [p.142].

As in all research, the experimenter wishes to employ a method which reduces the probability of Type I or false positive errors, yet at the same time avoids being unduly conservative and increasing the risk of Type II or false negative errors. However, while the controlled experiment can be utilized with some success with children in the middle childhood period (i.e., the period in which metalinguistic awareness emerges according to the second and third conceptual views) its use is considerably more problematic with the younger children in the age group specified in the first of the conceptual approaches noted previously. Specifically, as the age of the children being tested decreases, the use of the controlled experiment with its standard procedures increases the probability that at least their linguistic and perhaps their memorial and cognitive abilities will not be sufficient to meet the demands of the experimental situation. In essence, with younger children, the probability of Type II errors increases with the use of the controlled experiment, for it implicates variables which are irrelevant to the ability being investigated. For example, since very young children are still in the process of acquiring language, as well as developing their cognitive abilities, they may not understand a question put to them or a response they are required to give, even though they may possess the relevant metalinguistic ability. Consequently, their responses given under these conditions would result in an underestimate being made of their real ability.

Although it is not clear that a solution will be found for this problem of devising research methods which are suitable for use with very young children, it is important that the issue is pursued, for the theoretical and practical considerations noted previously. A twofold approach might be suggested. First, given that the controlled experiment provides the most valid (although most conservative) estimate of ability, its use should predominate over the more unsystematic methods, such as spontaneous speech records. However, since the latter can serve to provide a rich source of ideas and hypotheses, it should also be employed. Secondly, given that the analysis above suggests that the use of the controlled experiment increases the likelihood of Type II errors, researchers should devote more efforts to developing techniques which minimize extra-metalinguistic, particularly linguistic and cognitive, demands. We shall return to the latter point in the final section for a more detailed discussion, while noting here that methods which minimize irrelevant linguistic and cognitive demands have been used elsewhere to assess young children's reading [Calfee, 1977] and cognitive abilities [Miller, 1976] and there appears to be a strong case for devising analogous procedures for estimating metalinguistic abilities.

Multiple Representation of Metalinguistic Awareness

While the discussion in the previous section suggested that the three conceptual approaches differed concerning the onset of metalinguistic awareness, "onset" was discussed in that instance merely in terms of the first appearance of *some type* of metalinguistic ability. However, a second methodological issue which might be considered relates to the assessment of the variety of different competencies which are the result of the development of the ability to reflect upon and manipulate the structural features of language. Although it is unlikely that the field has yet identified all the different metalinguistic acquisitions, those presently delineated include awareness of phonemes, awareness of words, awareness of the structural representations of sentences, and awareness of interrelationships among propositions. An important conceptual (and methodological) issue related to these acquisitions is whether they emerge synchronously or whether they emerge in sequence.

Several viewpoints might be considered. One possibility is that the various metalinguistic abilities emerge synchronously although there may or may not be differences in the rate of their separate development (i.e., they may not co-terminate). Tunmer and Herriman [this volume] considered such a position in the second conceptual view that they discussed. According to that view, the different metalinguistic abilities which emerge in middle childhood each reflect a new kind of linguistic functioning. In turn, this acquisition ultimately depends upon the development of the child's metacognition; that is, his awareness of, or his ability to control his own intellectual processes. Given this common source, it is suggested that the different metalinguistic abilities may emerge synchronously, although their rate of development would differ as a function of the component skills unique to the particular tasks designed to measure each ability. It might be noted that a similar argument has been made in relation to the various abilities which are considered to comprise metacognition, including metamemory, meta-attention and metasocial cognition. Gleitman et al. [1972], for example, have suggested that the various "metas" may be closely related and may share a common period of emergence. In fact, they go so far as to suggest that this period would also encompass the appearance of metalinguistic awareness, a view which would seem to follow given the common source of the various metacognitive and metalinguistic abilities.

Alternatively, Flavell and Wellman [1977] have speculated that there may be an ordered sequence of emergence in the various "meta" abilities. According to these writers,

> it seems plausible that those psychological processes of self and others that tend to be relatively more external and therefore more accessible to perceptual inspection ought to become objects of knowledge earlier than those that are relatively less overt ... Under this argument, knowledge of, say, internal memorization and retrieval ("memory" in the narrower, conventional sense) ought in general to develop later than, say, metalanguage, since speech and writing are external and perceptible. [p.25]

While it is of some interest that Flavell and Wellman suggest that metalinguistic awareness may emerge prior to other metacognitive abilities, consideration might similarly be given to the possibility that the appearance of the various *metalinguistic* abilities is also characterized by asynchrony. That is, it might be argued that while most of these abilities are acquired primarily during the period of middle childhood (such that performances on tasks requiring these abilities tend to be positively correlated, when the effects of age are partialled out), some abilities are acquired prior to others. This suggestion then raises the issue of whether the asynchronous emergence and development of the various types of metalinguistic ability occurs in a particular and common sequence or whether the sequence depends more upon the unique experiences of particular individuals.

Arguing in favour of the former possibility, Rozin and Gleitman [1977] hypothesize that the order of emergence of the different metalinguistic abilities is a function of the *level* of linguistic representation being accessed, where phonological representation is the lowest level and semantic/pragmatic representation is the highest. According to this view, highly-processed, or "deep", representations are the easiest to access and report upon, while less-processed, or "surface", representations are more difficult. With regard to the emergence of the different metalinguistic abilities in children, Rozin and Gleitman claim that "the lower the level of the language feature, the later its accessibility to the language-learning child" [p.90]. That is, "the lower the level of linguistic organization called for, the more difficult it is for young children to respond to non-communicative linguistic activities ..." [p.93].

Again, another sequence of emerging metalinguistic abilities might be proposed based on an argument analogous to the Flavell and Wellman viewpoint noted above. It might be speculated, for example, that aspects of pragmatic awareness (e.g., being aware of the need to adjust one's speech to meet the abilities and experience of one's listener) might emerge before lexical and phonological awareness since the former are more in the public domain and subject to comments and criticisms from others.

Alternatively, a plausible developmental sequence of metalinguistic acquisitions might be proposed based on a conceptual analysis of the processes comprising each of the different metalinguistic abilities and a comparison of their relative levels of difficulty. As one aspect of pragmatic awareness, consider Markman's [1979] analysis of the component processes involved in tasks that require children to detect inconsistencies between the sentences of stories read to them. (More generally, of course, a similar analysis might be made of the processes implicated in the detection of inconsistencies in any communication addressed to the child.)

> As each sentence is read, the children must listen to, encode, and store the meanings or propositions that the sentence expresses. In order for two sentences to be seen as inconsistent, the sentence representations must be activated together in working memory. This in itself can be difficult, especially if time and information has passed between the two sentences. Establishing contiguity of the sentences in working memory is not sufficient; sentences must also be compared to notice their incompatibility. To notice an implicit contradiction, one must draw the relevant inferences as well. [p.653].

Clearly, this particular ability would appear to depend upon a number of processes and could not be evidenced until the component processes are themselves developed and can be engaged, concurrently or sequentially, as required. This ability would seem to require more component processes than would be involved in, for example, segmenting a word or syllable into its constituent phonemes. On these grounds, it might be speculated that the emergence of phonological awareness might precede that of the ability to detect propositional inconsistencies since the latter entails more, and more complex, processes than the former.

While the preceding discussion is highly speculative and it is possible that a more plausible developmental sequence of metalinguistic acquisitions might be proposed based on a fine analysis of the degree of similarity and interdependence of the different abilities, it is clear that this is an issue of some importance. On the one hand, whether or not the different abilities appear synchronously or asynchronously would obviously shed light on the basic nature of metalinguistic awareness and its relationship to the other "metas". On the other hand, the issue also carries potentially important practical implications, including the development of educational curricula appropriate for enhancing metalinguistic awareness [see Herriman and Myhill, this volume] and, perhaps, dependent abilities such as reading [see Tunmer and Bowey, this volume].

However, resolving whether the different metalinguistic abilities emerge synchronously or asynchronously depends upon locating a technique which would allow for a fair assessment of the issue. This is not a straightforward matter. To make this point clear, consider the issues facing the researcher. Drawing once again upon the distinction between task and procedure, the researcher must first decide upon each of the *tasks* which the child must be able to complete in order to be credited with competence in each of the different aspects of metalinguistic awareness. For example, in relation to phonemic segmentation, a child could be required to segment the first, second through to the nth phoneme in a word, to segment digraph and nondigraph words, to match phonemes in pairs of words, and to delete a phoneme and recombine the remaining elements in the words. The point to be made here is that while the aim of the researcher in selecting any one or more of these tasks is to gain a valid index of competence, the different tasks are likely to vary in the extent to which they provide a comprehensive assessment of a particular ability. And, this difference in comprehensiveness is likely to be associated with differences in task difficulty as a function of the different linguistic, analytic and memorial demands which are implicated in the separate tasks. Consequently, it might be anticipated that differences in estimates of the emergence of a *particular* ability might be obtained as a function of the specific tasks used, with the more comprehensive/difficult tasks producing more conservative estimates.

Acceptance of this point has important implications for the selection of the *group* of tasks to assess the synchrony-asynchrony issue for, ideally, the tasks required would be those that give equally conservative (or liberal) estimates of each ability. That is, the researcher would not want to conclude in favour of asynchronous development based on findings obtained on tasks which are differentially conservative in terms of the ability estimates provided. The problem for the researcher then, is to locate an equally conservative series of tasks. While there is no simple solution to this problem, one attempt at a solution would require the researcher to identify, for each metalinguistic ability, the range of tasks which would provide the least to the most comprehensive estimates of a particular ability. The researcher could then select, for each metalinguistic ability, a task (or group of tasks) from a similar point on the comprehensiveness dimension (i.e., low, moderate or high comprehensiveness).

While the preceding discussion focused on the problem created by different *task* demands, an analogous problem also inheres in the next step of developing *procedures* appropriate for assessing performance on the tasks related to each of the different metalinguistic abilities. Considering the example of phonological awareness again, a number of different procedures could be (and have been) used to assess phonemic segmentation ability. These procedures include, for example, the manipulation of coloured blocks [Calfee et al., 1973], tapping or clapping [e.g., Liberman et al., 1974], and sound matching [Wallach et al., 1977], to represent phonemes in spoken words. As the reader might anticipate, procedural differences of this sort must also contribute to differing estimates of segmentation ability being obtained due to the varying demands of the different procedures. For example, whereas Liberman et al. [1974] assessed segmentation ability by having children repeat a word and then tap on a table once for each phoneme, part of the procedure used by Calfee et al. [1973] involved the children being presented with syllables that consisted of two phonemes in different arrangements and being asked to manipulate coloured blocks to represent the relationship between the phonemes. The latter procedure would seem to require more representational, manipulative and memorial skills of the children than would the former procedure such that a more conservative estimate of segmentation ability might be expected if the latter procedure is used. Comparison of the findings obtained in these two studies provides some substantiation for this suggestion [see Nesdale et al., this volume].

As in the case of selecting the group of tasks, a solution to this procedure problem would again appear to entail the researcher in, first, identifying, for each of the metalinguistic abilities, the variety of procedures which could be used, and at what level of task comprehensiveness. Secondly, the procedures relevant to each ability would then need to be analyzed into their component processes. The researcher would then be able to exclude those procedures that implicate the most irrelevant linguistic and cognitive demands, yet retain suitable procedures to assess each level of task comprehensiveness. The procedures ultimately chosen would then depend upon the particular level of task comprehensiveness desired.

General Methodological Issues

While the preceding discussion has focused on methodological considerations in relation to two conceptual matters, there are a number of more general methodological issues regarding research on metalinguistic awareness in children. Each of these issues will be briefly discussed in turn. They are (1) the development of appropriate instructions; (2) the use of practice trials and feedback; and (3) the development of task-appropriate operations.

1. Development of Appropriate Instructions

It has long been accepted that there is an important need for instructions to research subjects to be clear, unambiguous, and specific, without being overly complex [Sutcliffe, 1972]. Given that these criteria are met, it is also recognized that a set of instructions can be used to serve several different goals including defining the purpose (or disguised purpose) of the experiment, describing the operations to be performed, developing a set, and motivating the subject [Sidowski and Lockard, 1966].

While it might reasonably be assumed that researchers typically attempt to facilitate some or all of these goals by developing a clear, unambiguous set of instructions, there is some doubt as to the extent to which this is achieved. For example, although the issue is not well researched, it has been reported in one of the studies which has addressed it, that despite the use of standard instructions, up to 50% of adult subjects were found not to have a basic understanding of the purpose of the study in which they had previously participated, nor of their task in the study [Cannell et al., 1977]. While the percentage of such noncomprehending subjects in any one study would presumably vary as a function of the type and complexity of the research and the care with which instructions are prepared, there is, nevertheless, a clear warning in these results for researchers assessing children's abilities. Specifically, if adults are prepared to participate as subjects despite their lack of a clear understanding of what is required of them, a similar type of response might be expected from children, particularly since their linguistic, memorial and comprehension abilites are less than those of adults.

There is some evidence consistent with this view. For example, Markman [1977] conducted several studies which were designed to assess the ability of 6-, 7-, and 8-year-old children to monitor their comprehension of a series of instructions (e.g., how to play a card game and perform a magic trick). Markman found that many of the subjects failed to comprehend the instructions and that they were not aware of this failure until they actually attempted to perform the operations. More pointedly, perhaps, Hughes and Grieve [1980] report findings which indicate that children will "make sense" of what is unambiguously nonsense. These researchers directed a series of bizarre questions (e.g., "Is red wider than yellow?") at 5- and 7-year-old children and found that the children generally responded and provided justifications for their responses (e.g., "Because it's got more colour"). Such findings suggest that children are quite prepared to provide responses to questions or instructions that they could not understand without developing or placing them in an idiosyncratic context.

The obvious implication of these results for researchers is that special care needs to be taken in order to develop instructions which are devoid of ambiguities, inconsistencies, gaps and irrelevancies. With young children, this is clearly not easy for the focus of the instructions is on entities (e.g., phonemes, words, sentences) with which the child is not particularly familiar, at least not as objects of observation. In addition, the child receives his instructions via a medium (i.e., language) over which he has a limited command (or, at least, not the command of an adult). As a result, instructions which are clear, unambiguous and direct to adults may be just the opposite to children because their understanding of the words used is not fully developed. In particular, they may not be familiar with the metalinguistic terms used (e.g., *word*, *sentence*, etc.).

In view of these issues, it appears that the solution regarding instructions is three-fold. First, the researcher should consider the development of a set of instructions as a necessary, time consuming and important component of the research activity. Secondly, a standard set of instructions should be developed so that the child's understanding of the task and procedures is not dependent on the wit of the experimenter at the particular time. Thirdly, the instructions should be explicitly pilot-tested for clarity rather than being judged acceptable simply on the basis of apparent clarity. Aside from checking the instructions with children in the target population, the researcher might also assess the reactions to his instructions of children who are a year or two older than those in the target population (i.e., given that the target population will typically be children in the middle childhood period). These children are more similar, linguistically and cognitively, to the target population than are adults, yet they should be sufficiently advanced to provide a useful commentary on the quality of the instructions developed.

2. Practice Trials and Feedback

A second procedural issue relevant to research on metalinguistic awareness concerns the use of practice trials and response feedback. The questions which arise here concern the purpose and number of practice trials to be used, the intent and type of feedback which should be given, and whether feedback should be given following each of the experimental trials.

Given that the child has been instructed as to what is required of him in the experiment, researchers frequently allow the child to engage in several *practice trials*. The point to be emphasized about such trials is that they should not facilitate the child's acquisition of an ability he previously did not possess. The use of practice trials should be limited to simply providing a practical illustration to the child of the operations required from him. To this end, the researcher needs to select several example stimuli which clearly represent the range of stimuli to be presented in the experiment and which allow the child to become familiar with the particular operations involved (e.g., tapping on a table, or re-arranging coloured counters, to represent the phonemes in one or more types of word). In addition, the number of practice trials given should be standardized and limited; practice trials should not continue to the point where the child eventually begins to reveal some success on the particular stimuli.

A recent study conducted by Tunmer and Nesdale [1982] to assess the effects of digraphs and pseudowords on phonemic segmentation in young children, provides an illustration of these procedures in operation. Based on a procedure used by Liberman et al. [1974], the phoneme segmentation task was presented in the form of a "tapping" game which required the child to tap once for each phoneme in a set of two and three phoneme words spoken to him. The researcher first demonstrated the response required by tapping out the number of phonemes in each item of a triad, such as *oo, boo,* and *boot*. The child was then asked to engage in the same behaviour. "Now, I want you to do it. Say *oo*. Good, Now, tap it. Say *boo*. Now tap it. Say *boot*. Now tap it." If the child made an error on any item, corrective feedback was provided and the item repeated. The items were then presented again, but in a different order. Since the experiment was concerned with relative performance on real words and pseudowords, the entire procedure was then repeated with a new training triad that involved a pseudoword, before the test trials commenced. Using this procedure, the child was thus exposed to an instance of each type of item in the test, in a limited number of trials.

As indicated in this example, practice trials also typically include *response feedback*. However, the provision of feedback to the child needs to be considered in relation to both the practice and experimental trials. In general, feedback has a twofold purpose, the two aspects being (or needing to be) differentially emphasised in the practice and test trials. One purpose of feedback is to shape the nature of the response required from the child. Although this directive goal should be pursued throughout each testing session, it should receive special emphasis during the practice trials. At this time, the feedback given should be used to indicate, for example, whether the child is attending to the relevant part of the stimulus, what is the appropriate amount of time to be spent on the required operations, and whether sufficient detail is being given in verbal responses. In addition, as part of this process, the child will be told whether his response is correct or not. Together, feedback on these points supplies information to the child as to what comprises adequate performance.

As noted above, such feedback, with the exception of outcome feedback (i.e., whether the child is correct or incorrect on a particular trial) should continue throughout the test trials. Although the researcher would not wish to confuse the child by abruptly terminating the outcome feedback from the practice to the test trials, he would also not want to shape his responses during the test trials by continuing the outcome feedback. As suggested by Miller [1976] in another context, one possible solution to this problem is to phase out the outcome feedback during the final practice trials.

The second purpose of feedback is to motivate the child. Feedback can have this motivational effect when it takes the form of the application of positive and negative evaluative statements following behaviour which is desired (e.g., attending, concentrating, giving complete responses) and undesired, respectively. This motivational function is particularly important with young children who are undoubtedly less willing than adults to work on repetitive, monotonous, uninteresting tasks [Orne, 1962]. (Brown [1978] in fact, provides an anecdote of some children who deliberately failed a recall test in order to play a more popular game.) However, while there is a clear case for researchers to develop effective feedback to enhance the children's motivation, particularly during the test trials when their interest and enthusiasm begin to flag, the little research available suggests that researchers do not use this feedback selectively nor effectively. For example, Cannell et al. [1977] report one study which examined the techniques employed in interviews with adult respondents in which it was found that primarily positive feedback ("OK", "all right", "that's good") was delivered, and that this feedback was delivered for both good (i.e., adequate) and poor (i.e., refusals, inadequate and "don't know") responses. It was claimed by the writers that the interviewers appeared to use the feedback "primarily to reassure the respondent (and themselves) by indicating receptivity and approval" [p.309]. If the behaviour of the interviewers in this study is any guide to the nature of interactions with children (a distinct possibility, given most researchers' concerns with establishing rapport with children), it would seem that there is a strong need for researchers to develop and clearly monitor their exchanges with children during testing sessions. If not, a potentially powerful technique for influencing behaviour will be wasted, or even worse, its indiscriminate use will confuse the children and/or reinforce inappropriate and undesirable behaviours.

3. Development of Task Appropriate Operations

Given that the tasks relating to a particular aspect of metalinguistic awareness have been selected, the researcher must then devise the operations which will be used to assess the child's abilities on these tasks. Considering the research which has been carried out so far, the child might be asked, for example, to move coloured counters; to tap or clap; to point; to give a verbal judgement of some sort; to pull a lever; to pronounce a sound, word or sentence; and to select or match pictures. While there are clearly a variety of operations which the child might be required to tackle in an experiment on metalinguistic awareness, these operations should satisfy several criteria.

First, the operations selected should be unambiguously appropriate to the particular task. If the task is to discriminate the words comprising a sentence, or to determine whether a consistent relationship obtains between a set of propositions, the operation selected should allow for no other interpretation. This is frequently not the case. To illustrate, consider the research conducted by Fox and Routh [1975] to assess 3- to 7-year-old children's ability to segment sentences into words. The experimenter first presented a sentence which was two to seven words in length after which the child was required to respond, giving "just a little bit of it .... For example, if I say 'Peter jumps', you would say, 'Peter'" [p.335]. The researchers reported that by age 4, virtually all of the children succeeded on this task. However, these results do not necessarily reveal an awareness of words as distinct linguistic units since the experimenter ceased asking for further segmentation when the child reached the point of producing word units. As a result, the finding may be an artifact of the procedure employed. In a similar fashion, an estimate of a child's grammatical awareness via his responses to the question, "Is the following sentence 'good' or 'silly'?", would also be ambiguous. As Gleitman et al. [1972] point out, "a number of tangled issues of truth, plausibility, meaningfulness and syntactic patterning enter into the interpretation of these findings" [p.156]. Research indicates that children from 2 to 3 years of age appear to judge sentences in terms of whether or not they are understood, while somewhat older children, aged 4 to 5 years, adopt a content criterion, rejecting many sentences that they understand but which say things they either do not believe or like. Since younger children tend not to evaluate sentences on grammatical grounds in a sentence acceptability task, this task may not provide an adequate assessment of children's grammatical awareness [see Tunmer and Grieve, this volume].

While the detection of the task inappropriateness of the operations in the preceding examples is relatively straightforward, this is clearly not always the case. For example, whereas the seemingly obvious operation to assess children's phonemic segmentation ability is to have them segment real words, research by Tunmer and Nesdale [1982] indicates that this may not be the case. In that study, children were required to tap the number of phonemes in digraph and nondigraph real words and pseudowords. Based on their findings, Tunmer and Nesdale concluded that tests comprised of real words,

> ... may provide inaccurate estimates of phonological awareness... particularly when used with beginning readers. Children who lack phonemic segmentation ability but read words by sight may employ a grapheme strategy ... [to segment the words] ... whereas children who have begun to master the grapheme-phoneme correspondence rules may resort to a spelling strategy. [p.309]

The non-obvious implication of their findings was that the most unambiguous estimate of a child's level of phonological awareness will be obtained when the test is comprised of nondigraph pseudowords. Together, these examples indicate that potential procedures should be closely analyzed, in relation to the specific tasks, in order to meet the appropriateness criterion.

Secondly, in order to reduce Type II errors, the operations devised should minimize the extra-metalinguistic demands placed upon the child. That is, the analytic, inferential and memorial demands of a particular procedure should be pared to a minimum. Illustrations of the way in which different demands are placed on children by different procedures are given throughout this volume. For example, Bowey and Tunmer suggest that the marked differences in word segmentation ability revealed in Tunmer and Bowey [1981] compared with previous studies may have been due to the additional strains imposed by the procedure used in the previous studies. In contrast to the relatively simple tapping procedure used by Tunmer and Bowey, the children in these studies were required to repeat sentence-long strings while simultaneously identifying their word boundaries. Similarly, in their review chapter in this volume on the development of grammatical awareness, Tunmer and

Grieve concluded that some of the discrepancies between studies may be due to the excessive memory demands of some procedures [e.g., Howe and Hillman, 1973] versus others [e.g., James and Miller, 1973]. Since this point was also taken up towards the end of the previous section, little further discussion is required. However, in terms of meeting the criterion of utilizing a procedure with minimal extra-linguistic requirements, the implication is that for any particular metalinguistic ability, the researcher should first identify the range of possible procedures which could be employed. The selection of a procedure for use should be based upon an analysis and comparison of the processes implicated in each procedure, with the procedure selected being that which exerts fewest extraneous demands.

Thirdly, despite the seeming incongruity, estimates of metalinguistic ability should be obtained using a procedure in which the subjects' verbal reports are reduced to a minimum. At present, studies of metalinguistic awareness typically require children to give verbal responses of one sort or another. Such verbal responses have several shortcomings as indices of metalinguistic abilities. Before detailing these shortcomings, however, it is worth noting that, despite the focus on children's metalinguistic *awareness*, the type of verbal response elicited in this research does not occasion the same criticisms as have been directed at the discredited method of introspection [Boring, 1953] and its contemporary elaborations [Cavanaugh and Perlmutter, 1982; Nisbett and Wilson, 1977]. With regard to the contemporary use of introspective techniques, there has recently been an upsurge of interest in the development of metacognition, especially metamemory [see reviews by Brown, 1980; Cavanaugh and Perlmutter, 1982; Flavell, 1978; Flavell and Wellman, 1977]. Consistent with the systematic experimental introspection conducted by Kulpe in Wurzburg [Peters, 1962], the focus of this research has been on *how* as well as *what* the child remembers in a particular memory task. That is, the research involves a verbal self-report methodology. As Cavanaugh and Perlmutter [1982] point out, however, this research is susceptible to the two most serious and perhaps unresolvable criticisms which were originally directed at systematic introspection. These criticisms relate to the "... [limited] accessibility of cognitive processes for introspective analysis and [the ambiguous] veridicality and completeness of a verbal report" [p. 16].

In contrast, the typical procedure in metalinguistic awareness research is more analogous to that involved in a psychophysics than a metamemory experiment. That is, the subjects are not required to introspect and report on the processes they engage in to produce a solution but, rather, are simply required to indicate what solution they reached. Moreover, the response required in metalinguistic research is given in relation to an event which is capable of verification. For example, subjects might be asked to listen to a word or a sentence and then to indicate how many phonemes comprise the word or how many words comprise the sentence. In each case, the response required has an objective referent. Unverifiable verbal self-reports such as are obtained in metamemory research have not been the focus of attention in metalinguistic research.

Nevertheless, it needs to be recognized that much of the research conducted on children's metalinguistic awareness is reliant on verbal responses and data obtained in this manner are subject to error. For example, verbal reports are subject to a variety of other influences such as evaluation apprehension, social desirability and demand characteristics [Smith and Miller, 1978]. In addition, a child who is metalinguistically aware could fail to be credited with this ability because of a difficulty in articulation, a failure to remember at the time that the verbal report is required, or because he did not understand the question put to him.

In view of these limitations of verbal reports, and because the requirement should be for the minimum necessary evidence that a child is metalinguistically aware, there are good grounds for minimizing the verbal component of experiments on metalinguistic awareness. On this point, it is noteworthy that several researchers have employed so-called non-verbal methods with some success in assessing Piagetian concepts such as transitivity, conservation and seriation [see Miller, 1976]. What should perhaps be emphasized about these methods is that while they are not literally non-verbal, all have at least reduced the typical verbal requirement, "especially with respect to the critical elements of the assessment procedure, that is, the form in which the criterial question is posed and the manner in which the child is required to respond" [Miller, 1976, p.406]. Given the potential influence of the verbal component on metalinguistic ability estimates, particularly with younger children, the development of analogous non-verbal procedures for assessing language awareness should be a prime concern. This is not meant to imply that this point has not already been recognized by some researchers. Earlier we noted the procedure of using counters or taps to represent phonemes in words and words in sentences [e.g., by Ehri and Wilce, 1980; Liberman et al., 1974; Tunmer and Bowey, 1981] as well as the phonemic segmentation procedure used by Calfee et al. [1973] in which the children were required to manipulate coloured blocks in order to represent the relationship between the phonemes in words. Such procedures as these clearly reduce the linguistic demands placed on the child, although care must also be taken that other non-linguistic (e.g., memory) demands are not increased as a result.

Finally, it should be emphasised that the suggestion to increase the use of non-verbal techniques does not imply that verbal responses should never be obtained from the child. Clearly, in some situations the gain from having a child give a non-verbal rather than a verbal response would be difficult to establish (e.g., the difference between having a child press a red button versus saying, "red"). Indeed, at times it may be counterproductive. More importantly, however, research has shown that explanations which accompany verbal or non-verbal responses can provide a useful source of information regarding the basis of the child's response. For example, in a study conducted by Tunmer et al. [in press] to assess children's ability to detect propositional inconsistencies, 5-, 6- and 7-year-old children first indicated whether a particular story which was read to them was "a good story which made sense or a silly story which did not make sense" and were then asked why they thought this was so.

While the judgement data obtained in this experiment indicated that the children performed better on some types of stories than others, analysis of the explanations given for their judgements suggested some reasons for these differences in performance. For example, the results indicated that the 5-year-old children performed considerably better on stories in which propositions were inconsistent with an explicitly rather than an implicitly stated main premise. Analysis of the accompanying explanations suggested that the poorer performance on implicit stories may have been due to the children's tendency to judge these stories in terms of the truth of individual sentences rather than the inconsistency between propositions. Clearly, these findings suggest that explanations accompanying judgements can assist interpretations of particular patterns of performance. Of course, this in no way implies that a correct judgement-plus-explanation criterion should be applied in this research. Consistent with other writers [e.g., Brainerd, 1973], judgements rather than judgements-plus-explanations should comprise the dependent measure in such research in order to reduce the probability of Type II errors resulting from articulation problems.

Concluding Remarks

The primary aim of this chapter has been to examine some of the methodological issues which need to be addressed by researchers interested in the development of metalinguistic awareness in children. In general, these issues arise from the limited linguistic, memorial and cognitive abilities of young children, limitations which impose special demands on researchers attempting to employ controlled experiments for data collection. In tackling these issues, we have emphasized the need to undertake both task and procedure analyses prior to selecting tests and conducting research since both task and procedure contribute to the comprehensiveness and difficulty level of a specific assessment method. The particular importance of these analyses was revealed in relation to, first, the different theoretical positions regarding metalinguistic awareness and, secondly, the multifaceted nature of this acquisition.

In addition to these issues, we have considered three of the essential components of research, each of which has an important bearing on the collection of reliable and valid data in research with young children. These issues include the preparation of appropriate instructions, the use of practice trials and feedback, and the development of task-appropriate procedures. While there is perhaps a general tendency to focus on the latter issue as the most important and to ignore the remaining two, our analysis indicates that all three are important and that each can exert a sizeable influence on the results obtained. Beyond emphasizing the need for careful planning to control for these influences, however, two final points can be made in relation to the goal of establishing valid methods for assessing metalinguistic awareness. Given that the process of method refinement is now well underway, there is a clear need for, first, replication studies to establish data reliability and, secondly, longitudinal studies to provide uncontaminated trend data.

Part 2

Emergence of Metalinguistic Abilities

2.1 Phonological Awareness in Children

Andrew R. Nesdale, Michael L. Herriman, and William E. Tunmer

In examining the young child's awareness of language, an appropriate place to start would seem to be with his awareness of the most elementary units of language, that is, with the child's awareness of the phonological units. At this level the focus is on the extent to which the young child can both segment the spoken word into its component phonological units and synthesize these units to produce a word. While such a starting place might be considered to have an inherent appropriateness since the phoneme appears to be the most elementary unit of language, it has also become increasingly apparent in recent years that the child's acquisition of phonological awareness is a crucially important achievement. Specifically, a series of recent studies has begun to articulate a significant role for phonological awareness in facilitating the child's learning to read. On these grounds alone, it is clear that an understanding of this aspect of metalinguistic awareness is of no small importance.

The purpose of the present chapter is to provide a brief overview of research which has focused on phonological awareness, particularly the young child's ability to discriminate the phonological units of language. Special emphasis will be given to findings bearing on such questions as; at what age do children become phonologically aware, is there a developmental trend in this achievement, what does the development of phonological awareness depend upon, and can this aspect of language awareness be trained? By way of introduction, we will briefly examine the relationship of phonological awareness to reading. In addition, some consideration will be given to the characteristics of the spoken language which contribute to the child's difficulties in phonemic segmentation and to the experimenter's difficulties in assessing the child's level of awareness.

Phonological Awareness and Reading

In considering the relationship between phonological awareness and reading, we might start with the question, "What does a child need to know in order to read an alphabetic language properly?" Clearly, as Golinkoff [1978] and others have noted, the child first needs to be able to speak the language and to be able to discriminate among the alphabet's orthographic units. Given these skills, there are then at least two common strategies he might follow in learning to read, each requiring one or more additional abilities.

One strategy, the "look-and-say" or "whole word" approach, simply requires the child to learn visual-auditory correspondences at the level of words. Thus, each new word is learned as a unit (e.g., the visual form DOG is pronounced *dog*). The advantage of this approach is that given an adequate memory, the child can fairly quickly learn a large enough vocabulary of recognizable visual forms such that he can progress to reading texts which use these words. It is at this point, however, that the main deficiency of the holistic strategy is revealed: it does not permit the child to progress independently since he is unable to read words not previously encountered in print.

In contrast, a more analytic approach which involves relating the orthographic components of the written word (in the visual form D-O-G) to the *segmental* structure of the spoken word (i.e., the spoken *dog*) would seem to allow for such progress. As Liberman et al. [1977] point out, "given a word that is already in his lexicon, the child can read it without specific instruction although he has never before seen it in print, or given a new word which he has never before heard or seen, the child can closely approximate its spoken form and hold that until its meaning can be inferred from the context or discovered later by his asking someone about it" [p.209]. While the analytical strategy would thus appear to have distinct advantages, it is claimed that such an approach requires an additional special ability. Specifically, aside from being able to produce the appropriate sounds for individual letters, the child has to become aware of the fact that speech consists of phonemic units; he has to be able to analyze and synthesize the phonemic units that comprise spoken words [Golinkoff, 1978; Liberman et al., 1977]. Knowing the phonemic units in the spoken word, the child is then able, when confronted with the printed word, to accurately map the latter onto the former. Liberman et al. [1977] make this point especially clear with an example in which the child is confronted with the printed form *bag* and all he knows is the individual letter-sound correspondences. In this case, the child will sound out the word as *buhaguh*, a nonsense trisyllable containing five segments and not as *bag*, a monosyllable which has only three segments. As Liberman et al. point out, if the child is to map the printed *bag* onto the spoken *bag* which is already in his lexicon, he must be aware that the spoken word also has three segments.

Golinkoff [1978] has recently reviewed many studies which have examined the relationship between phonological awareness and reading ability. These include correlational studies [e.g., Rosner and Simon, 1971; Calfee et al., 1973; Liberman et al., 1974], experimental studies [e.g., Goldstein, 1976; Fox and Routh, 1976] and training studies [e.g., Rosner, 1971; Elkonin, 1973; Wallach and Wallach, 1976]. We shall subsequently examine aspects of some of these studies that pertain to phonological awareness in greater detail, and merely note here that there appears to be some support for the claim that phonological awareness is critically implicated in the acquisition of reading skill. As Golinkoff [1978] concludes,

> ...phonemic awareness skills -- both analysis and synthesis -- have been shown in a number of studies to be predictive of early and extended reading achievement. In fact, some studies suggest that phonemic analysis skills may be necessary for success with early reading instruction. For the child who may not have naturally acquired such skills...the literature suggests that their reading skills may suffer...If a child has received some type of phonemic awareness training, the literature indicates that the child's reading achievement is likely to be boosted significantly above where it would have been without training. [p.38]

The Difficulties in Acquiring Phonological Awareness

While the accumulating evidence points to the importance of the child's being aware of the phonological properties of words, it is important to note the difficulties the child faces in acquiring this awareness. What is entailed is that the child must first be able to attend to the spoken word as an object and must then realize that it is comprised of insolable segments or phonemes. While the evidence suggests that most children up to the age of 5 years do not (or cannot) attend to the spoken word [Rozin et al., 1974], work by Liberman and his colleagues indicates that when children do begin to attend to the spoken word, they are faced with a particularly difficult analysis task [Liberman et al., 1967]. In essence, the problem is that there is no simple one-to-one correspondence between the segments of a speech signal and the phonemes we hear; that is, there is often no acoustic criterion to indicate where one phoneme ends in a word and another starts. In experiments involving attempts to synthesize speech sounds using machine-readable spectrograms, Liberman and his colleagues demonstrated that at any one time at least two bits of information about phonemic content are being transmitted in the spoken word. This phenomenon is referred to as "parallel transmission" and is illustrated in Fig. 1 in the phonemic overlap present in producing the word *bag*. The middle part of *bag* not only carries information necessary for identifying the vowel, but also information necessary for identifying the initial and final consonants.

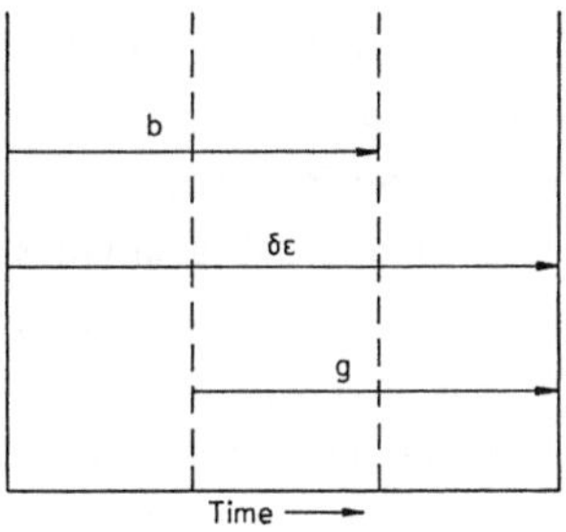

Fig. 1. Parallel transmission of the word *bag* [after Liberman, 1970]

This finding suggests that phonemes are highly abstract in that what had been thought to correspond to a phoneme is not a simple sound on the perceptual level. Yet, as Liberman and others have noted, listeners can separate the spoken word into distinct segments. While we do not intend to pursue the question of how speech sounds are perceived, suffice it to say that the sound of the phoneme depends not on its individual shape but on its relationship with other parts of the utterance [Foss and Hakes, 1978]. What should be clear however, is that the abstractness of the phoneme increases the child's difficulty in phonemically segmenting an utterance. It is simply not possible to segment the utterance such that each acoustic unit corresponds to one phoneme. As was illustrated in the example above, phonemes more appropriately correspond to syllables if they are rendered in their acoustic counterpart (e.g., *buh-ah-guh* for *bag*). To segment an utterance into its constituent phonemes, the child needs to be aware that such units are only "imprecise and abstract analogues of the way these phonemes sound in the word" [Golinkoff, 1978, p.24].

The Assessment of Phonological Awareness

Aside from emphasising the importance of the acquisition of phonological awareness, the preceding discussion perhaps also begins to suggest the complexities involved in assessing this aspect of metalinguistic awareness.

The complexities depend mainly upon the two related issues of *what* do we assess and *how* do we assess it?

While we might agree that being phonologically aware implies that the individual knows that the spoken word can be segmented into phonemic units, that he can distinguish between these units, and that he can synthesize such units into words, there is considerable flexibility in exactly what we can require the child to do in order to credit him with a certain level of phonological awareness. For example, in assessing the ability to segment, researchers have asked children to isolate the first, second, through to the last phoneme in spoken words; to match phonemes in pairs of words; and to pronounce words after deleting individual phonemes. While performance on any one of these tasks might well be considered to say something about a child's phonological awareness, it is also certainly plausible that such tasks might demand different levels of segmentation ability in the child. In fact, Golinkoff [1978] speculates that "recognizing the presence or absence of a unit should be easier than adding or deleting the element itself. Performing a deletion and recombining the remaining elements should be easier than performing the deletion and replacing the deleted element with another element" [p.26]. If such a view is substantiated, the clear implication is that considerable caution must be exercised in comparing the results of specific studies since the child's assessed level of phonological awareness will depend greatly on the task. That is, it will depend on *what* he is asked to do.

The additional consideration which compounds the problem of assessment is that the specific procedure (i.e., the *how*) used to assess awareness in these tasks might also influence the child's performance. Clearly, whatever the task we require the child to undertake, our concern is to utilize a procedure which provides a relatively *pure* or unambiguous measure of the child's ability on this task. However, when the relevant literature is considered, it appears that not one universally accepted technique is presently available which meets this requirement.

For example, children have been asked to isolate phonemes by clapping or tapping [e.g., Liberman et al., 1974], by representing them with different coloured blocks [e.g., Goldstein, 1976], by indicating which of two pictures representing different words begins with a given initial sound [Wallach et al., 1977] and by simply pronouncing only "a little bit" of a given word [Fox and Routh, 1975]. Just as with the range of tasks used, it would seem that such procedures might also vary in difficulty level, largely as a result of the number of operations the child is asked to perform. Thus, compared with the "little bit" technique of Fox and Routh [1975], children in the Liberman et al. [1974] study were required to perform the two tasks of repeating a word and tapping the phonemic segments. If the latter procedure represents an increment in difficulty level, we might expect that phonemic awareness assessed via this procedure would be reduced.

Furthermore, regardless of whether the specific assessment technique influences the child's performance, it might be anticipated that the amount of explicit or implicit training could contribute to performance, as would the extent to which feedback is given. Again, we find differences between researchers in the procedures adopted. For example, compared with studies which require subjects merely to segment one- or two-syllable words [e.g., Goldstein, 1976], children in other studies have been asked to segment a sentence into words, then words into syllables and syllables into sounds [e.g., Fox and Routh, 1975]. Although the training was not explicit, it is clear that children in the latter task might be undergoing implicit training which could influence their level of performance at the level of phonemic segmentation.

In sum, the assessment of phonological awareness is not clear-cut; potentially at least, both what we ask a child to do and how he is required to do it will contribute to his assessed level of performance. We shall now consider in more detail some of the studies which have examined young children's phonological awareness. In our treatment, we will focus on children's phonemic segmentation abilities rather than on both segmentation and synthesis. Although the acquisition of the latter skill is clearly an important achievement for the young child, it has not received as much attention as segmentation. Moreover, while research has been reported which suggests that synthesis may be both independent of, and of easier acquisition than, segmentation [e.g., Goldstein, 1976], other research suggests that the acquisition of synthesis skills may depend on the availability of some minimum level of segmentation ability [Fox and Routh, 1976].

Phonemic Segmentation

It must first be recognized that the potentially influential factors considered in the preceding discussion have not been systematically examined in research studies. Consequently, we can only provide provisional answers to some of the questions raised at the outset of this discussion (at what age do children begin to display phonological awareness, is there an age-related trend in this acquisition, etc.).

A distinction which needs to be made clear at the outset concerns that between speech discrimination (i.e., the ability to distinguish individual sounds containing different phonemes) and phonemic awareness (i.e., the ability to isolate the phonological units of a spoken word). The point to be made here is that speech discrimination skills are independent of and precede phonemic awareness skills. Studies of speech perception have shown that infants as young as one month can discriminate between small differences in phonemic features [e.g., Eimas et al., 1971; Eimas, 1975; Morse, 1972]. For example, when presented with the sounds /ba/ and /pa/, infants discriminate between these two syllables which involve differences in the onset time of voicing in the initial consonant sound. Infants of this age, however, are not aware of the phonological distinctions involved [Clark and Clark, 1977].

A difference in performance on discrimination and awareness tasks has also been reported by Wallach et al. [1977]. In one experiment, pre-reading kindergarten children from disadvantaged and from middle-class backgrounds were shown pairs of pictures with the subject of each differing only in the initial phoneme of its name (e.g., *jail* and *whale*). When the names were read, children were to point to the relevant picture (i.e., a speech discrimination task). Wallach et al. [1977] reported that the children got nearly all items correct with no significant differences between socio-economic class. In contrast, in a task used to assess phonemic awareness in which the children were shown pairs of pictures with maximally contrasting subject names (e.g., *man* and *house*) and were asked to indicate the picture which started with a specific sound (e.g., /m/), significant differences were obtained. Most of the middle-class children scored at or near the maximum possible, while a majority of the disadvantaged children were correct on 50% or less of the trials. This result, which was essentially replicated in two other experiments, indicates that lack of phonemic awareness does not imply a lack of speech discrimination. The two are independent skills.

Leaving aside for the moment the background differences in the results, the finding that at least the middle-class children performed at a high level on the phonemic awareness task provides a convenient starting point for our consideration of the findings which have been obtained on such tasks. The Wallach et al. finding, together with a similar result by McNeil and Stone [1965], are of particular interest for it appears to conflict with other claims regarding young children's phonemic segmentation abilities. Bruce [1964], for example, concluded from his experiment that the phonemic analysis of words could not be accomplished by children with mental ages below 7 years. The mean percent correct responses of children with mental ages of 5, 6 and 7 on his test were 0%, 6% and 29%, respectively.

However, it is perhaps at this point that the previous discussion on task differences and their impact on assessing phonological awareness begins to take substance. Compared with the Wallach et al. [1977] task which required children simply to recognize that a word begins with a given sound, Bruce's [1964] task may have been markedly more difficult. Specifically, to demonstrate segmentation ability in the Bruce experiment, children were required to say what word would be left if a particular letter sound (first, middle or last) were to be taken away from the test word (e.g., (h)ill). As Fox and Routh [1975] have pointed out, this is a difficult task which necessitates at least two operations. For example, in the case of a medial sound detection (e.g., lo(s)t) the task requires both analysis and synthesis rather than analysis alone.

In fact, in recognizing the potential impact of task differences on segmentation performance, Rosner and Simon [1971] devised a series of seven analysis tasks which they considered differed systematically in difficulty level. The tasks were similar to Bruce's in that they required the children to omit particular segments from words. Of the seven tasks used by Rosner and Simon, four are relevant to our discussion. These are tasks III through VI:

III. Omission of the final consonant of a one-syllable word (e.g., (bel(t)).
IV. Omission of the initial consonant of a one-syllable word (e.g., (l)end).
V. Omission of the first consonant of a consonant blend (e.g., (b)lock).
VI. Omission of a medial consonant (e.g., de(s)k).

The tasks were administered to kindergarten through to sixth grade children in two sections with a mixed order of items from the easier tasks (including Tasks III-V) presented in the first section and a mixed order of items from the harder tasks (including Task VI) presented in the second. The phonemic analysis results obtained by Rosner and Simon [1971] are presented in Table 1.

As might be expected, these results reveal that on each task there is a progressive increase in performance from kindergarten through to sixth

Table 1. Mean percent correct responses to word types by grade level [from Rosner and Simon, 1971]

	Grade Level						
Task	K	1	2	3	4	5	6
III	20.0	81.8	80.8	91.6	94.9	94.5	94.0
IV	7.0	70.2	86.9	93.7	94.8	97.6	97.0
V	6.3	44.3	42.9	60.8	63.8	71.1	74.7
VI	0.5	22.6	33.5	53.1	56.9	62.1	74.1

grade. In addition, however, there are several other features of these findings which are of particular relevance to our discussion.

Consistent with Rosner and Simon's hypothesis it is notable that performance is clearly task-dependent; there is a decline in performance from Tasks III-VI, within each grade. The second feature to note is that the kindergarten children performed poorly on all tasks with the mean percent correct response on the easiest task (Task III) being only 20%. Finally, there is a noticeable improvement in performance on all tasks from kindergarten to first grade. In fact, the results indicate that at least on Tasks III and IV, first grade children showed a high level of performance (mean percent correct responses of 81.8 and 70.2 respectively).

Two issues are raised by the latter observation. First, the relatively large gap between the performance of kindergarten and first grade groups on each task (particularly compared with the relatively small gaps between other pairs of grades) begs the question of why such a difference might occur. Several possibilities can be entertained. On the one hand, it might simply be due to a below average sample of kindergarten children or an above average sample of first grade children. On the other hand, and perhaps more plausibly, it may be that the increment reflects some aspect of the children's school experience. One possibility is that the result demonstrates the effect of exposure to reading instruction. That is, while we have observed that phonological awareness skills contribute to success in reading acquisition, there is also some evidence that reading instruction itself may indirectly influence the development of segmentation skills. Bruce [1964], for example, found that children from a school in which the method of reading instruction emphasized analysis skills showed more phonemic awareness than did children from a school where the reading instruction emphasized whole-word training.

The second issue raised by Rosner and Simon's results concerns the age at which young children can display phonemic awareness skills. Clearly, since the first grade children in this study did perform so well on at least Tasks III and IV, Bruce's [1964] conclusion that children with mental ages of less than 7 years cannot show phonemic awareness, needs to be qualified. At the least, the extent of awareness credited to a young child depends very much on the task which he is required to undertake. As we shall see, this point receives further substantiation from other studies.

Goldstein [1976] reported that children as young as 4 years achieved mean scores ranging between 17% and 46% on four phonemic analysis tests administered in two sessions. However, once again, the results are undoubtedly influenced by both task and procedure. On the one hand, the children's task was the relatively common one of isolating the component units in one-syllable two-phoneme, and one-syllable three-phoneme words by pronouncing the segments (e.g., "tuh-ee" for *tea*). On the other hand, the procedure involved the child participating in four sessions with the two phonemic analysis sessions following two sessions in which the children had practice in segmenting the syllables in two- and three-syllable words. Furthermore, prior to the analysis trials in each of the four sessions, the child was given a series of synthesis trials where the experimenter pronounced the segmented word (e.g., "tuh-ee") and the child had to say the synthesized word. Finally, in the trials in each session, half the words to be segmented by the child had previously been presented in segmented form by the experimenter for the child to synthesize, while half were new words. Clearly, aside from the simpler segmentation task used, the obtained findings might be presumed to reflect a strong training component. In partial support of this view, the analysis demonstrated a reduced performance on the words to which the children had not been previously exposed.

A similar evaluation can also be directed at a phonemic analysis study conducted by Fox and Routh [1975]. In view of what they saw as the unnecessarily difficult tasks and procedures used in other phonemic analysis studies, these researchers set out with the explicit aim of utilizing a task and procedure which would have "fewer extraneous cognitive requirements than any of the (previous) studies" [p.333]. Thus, following the experimenter's oral presentation of two- or three-phoneme syllables, the child was required to respond with, "just a little bit of what I say...For example, if I say 'Pete', you would say 'pe'," [p.336]. The results indicated a markedly improved performance compared with previous studies. Mean percent correct for 3-year-olds was about 28% which rose to 70% for 4-year-olds, 86% for 5-year-olds and approximately 93% for 6- and 7-year-olds. However, inspection of their procedure reveals that the phonemic segmentation trials were preceded by sentence-into-word and word-into-syllable trials; the child was apparently given continuous feedback and correction trials, and he was rewarded with raisins for correct responses. Any one or all of these factors might have contributed to the enhanced performance. More importantly, the results obtained using this procedure may not indicate that children are aware of *phonemes* as linguistic units distinct from phrases, words and syllables, since the experimenter continued asking for further segmentation until sound units were produced, and only then stopping.

Although to this point, the seemingly obvious conclusion to draw is that a child's segmentation ability is both task- and procedure-dependent, one further crucially important point needs to be made in relation to these studies. Specifically, it is arguable whether the procedures employed in any of these studies provide a *pure* test of children's phonemic segmentation abilities. Recall that in these studies, children were required to match a given letter sound [Wallach et al., 1977], to pronounce a word after a presented letter sound was elided [Bruce, 1964; Rosner and Simon, 1971], to utter the phonemic segments of a word [Goldstein, 1976], or to pronounce "a little bit" of a word [Fox and Routh, 1975]. Such procedures involve the child matching, speaking or deleting phonemes which are embedded in syllables containing two phonemic segments. But as we have pointed out previously, such segments do not correspond to the phonemic segments that occur in the spoken word. In this sense, to reiterate Golinkoff's [1978] point, the child has been asked in these studies to recognize or manipulate units that are only "imprecise and abstract analogues of the...phonemes in the word" [p.24]. Clearly, the recognition of this point clouds any conclusions we might want to draw from these studies concerning children's phonemic segmentation abilities. The implication is that the procedures used must be carefully designed to ensure that they do assess *phonemic* segmentation. When we consider the remaining studies which have addressed the issue of phonemic segmentation, few appear to satisfy this criterion. Moreover, studies which have attempted to satisfy this requirement have revealed other limitations and for the results obtained indicate that it serves to increment the difficulty level of the test used.

Calfee [1977] for example, developed a test designed to measure a child's ability to strip off the initial consonant portion of CVC or CCVC words and pronounce the remaining VC segments (e.g., *soap* - *oap*). In the training trials, it was ensured that the child was given every opportunity to comprehend the task. The child was first taught the response words (i.e., *eat, eyes, ache*) and provided with a set of cards depicting the acts or objects associated with each response word. The experimenter then told the child "If I say *greet*, you should say *eat*, if I say *ties*, you should say *eyes*," etc. Thus the procedure did not involve *verbally* asking the child to delete a phoneme from a word, which would have required embedding the phoneme in a syllable containing *two* phonemes (e.g., "Say *gate* but without the *guh* sound"). Kindergarten children who were administered this task scored a mean of 90% correct on the training trials, and when presented with new stimulus and response words to test generalization of segmentation skill, their mean performance was 70% correct.

Although Calfee's test is a considerable improvement over other tests of phonological awareness, a possible limitation is that it does not provide a measure of the child's ability to segment *successive* phonemes in spoken words. Stripping off the first phoneme may not be as difficult as segmenting the middle vowel, for example, which, as noted earlier, not only carries information necessary for identifying the vowel, but also information necessary for identifying the preceding and following consonants. That is, the parallel transmission of phonemic content results in a "stacking up" of information such that phonemes which are surrounded by other phonemes may be more difficult to segment.

An alternative procedure which has been developed requires children to use coloured blocks, counters or taps to represent successive phonemes in spoken words [e.g., Calfee et al., 1973; Ehri and Wilce, 1980; Liberman et al., 1974]. In a study by Calfee et al. [1973], for example, estimates of the phonemic awareness of kindergarten through twelfth grade children were obtained by a procedure involving the arrangement of coloured blocks. Specifically, in one test the child was presented with syllables comprising the two phonemes /i/ and /p/ in different arrangements (e.g., /pi/, /ip/, /pip/) and was required to manipulate coloured blocks to represent the relationship between the phonemes. In the second test, more complex syllables were presented (e.g., "if that says /ips/ show me /isp/") and again the child had to represent the relationships with the coloured blocks. The results indicated that even high ability first grade children achieved, on average, only about 40% correct responses on the first test and only 30% on the second test. The high ability kindergarten and second grade children scored 8% and 66%, respectively, on the first test and 18% and 62%, respectively, on the second test.

While the procedure used in this experiment ensured that the children's phonemic segmentation abilities were assessed, their poorer performance compared with other research suggests that the test (i.e., both task and procedure) was considerably more difficult. First, considering that there was no constant relationship established between specific colour and specific phoneme, it is perhaps not surprising that young children performed poorly on the tests. In addition, the second test required both phonemic analysis and synthesis. Perhaps the clearest indication of the level of difficulty of the test is revealed in the finding that high ability sixth grade children only scored 70% and 72% respectively on the two tests.

A tapping procedure which was used by Liberman et al. [1974] appears to minimize some of these difficulties. In that study, preschool, kindergarten, and first grade children were asked to isolate the phonemic segments of 42 randomly presented one-, two- and three-segment syllables. The children were required to repeat a spoken word (or sound, in the one segment case, e.g., /e/ as in *bet*) and to indicate by tapping a small wooden dowel on a table, the number of segments in the word or sound. Using a criterion of six consecutive correct test trials the data revealed an increase in performance with grade. Whereas none of the preschool children and 17% of the kindergarten children reached criterion, 70% of the first grade children showed phonemic segmentation.

However, as Tunmer and Nesdale [1982] have pointed out, a possible difficulty with a test of this sort, is that it may confound the segmentation of phonemes with the segmentation of graphemes. As noted earlier, children in the first stage of reading acquisition rely primarily on a strategy in which words are learned by sight. Venezky [1976] argues that sight word recognition requires attending to both letter orientation and letter order. Beginning readers who lack phonemic segmentation skills but read words by sight may therefore resort to a "grapheme" strategy in the phoneme tapping task, tapping once for each *letter* in a given word. Since most short, high-frequency words contain as many letters as phonemes, this strategy would lead to artifactually high scores among beginning readers, which might contribute to the sharp increase in performance shown by the first graders in the Liberman et al. [1974] study. If some first graders were, in fact, employing a grapheme strategy, there should have been a tendency for these children to "overshoot" on familiar words which contained digraphs, or letter pairs which represent single phonemes (e.g., *sh, th, oa, oo*). That is, such children would have incorrectly tapped four times on a word like *book*, instead of three. Since it could not be determined whether this actually occurred in the study by Liberman et al. because item scores were not reported, Tunmer and Nesdale [1982] conducted a study to explore this issue. Kindergarten and first-grade children were asked to perform a modified version of the phoneme tapping task in which 12 of the 24 test items were high-frequency real words, and 12 were pronounceable pseudowords. Half of the words in each of these groups contained digraphs (or would, if they were real words), while the other half did not. Tunmer and Nesdale hypothesized that if some beginning readers do rely on a grapheme strategy to segment words, higher scores would be observed on the words not containing digraphs than on those that do, since the children would tend to overshoot on the latter. However, it was expected that this difference would only occur with real words, since, by definition, pseudowords have not been seen before, and therefore cannot be learned by sight.

Analysis of the results obtained indicated that the kindergarten children's scores distributed bimodally with one quarter of the children scoring above 70% correct while the remainder performed very poorly. In contrast, analysis of the first grade children's results revealed three clear patterns of performance. As indicated in Table 2, the first response group was comprised of children who were generally successful in segmenting all word

Table 2. Mean number of correct responses as a function of word type, digraph type and group, and mean percentages of total errors that were overshoot errrors on digraph words [from Tunmer and Nesdale, 1982]

		Real			Pseudo		
		Nondigraph	Digraph		Nondigraph	Digraph	
Group	N	Correct	Correct	Over-Shoot (%)	Correct	Correct	Over-Shoot (%)
1	7	5.57	4.57	67.8	5.71	5.00	50.00
2	7	4.71	1.14	65.4	4.71	1.71	72.6
3	6	3.00	.83	33.8	2.00	2.83	8.3

types, regardless of whether the words were real or pseudo, or contained digraphs or nondigraphs.

The second response group was comprised of children who were considerably more successful in segmenting nondigraph rather than digraph words, regardless of whether they were real or pseudo. This group is of particular interest for their performance on nondigraph pseudowords suggested they could segment phonemically, whereas their performance on digraph words indicated otherwise. As indicated in Table 2, these children showed a greater tendency to overshoot than undershoot on both the real and pseudowords containing digraphs. It would appear that they used their knowledge of the grapheme-phoneme correspondence rules to generate a graphemic representation of the words, and then tapped out the graphemes rather than the phonemes. That is, the children relied on a spelling strategy to perform the task.

The final response group contained children who performed poorly on both the digraph and nondigraph pseudowords and real words. Nevertheless, their performance decreased from real to pseudo nondigraph words and increased from real to pseudo digraph words. These results, together with the overshoot errors, suggested that not only were these children unable to segment phonemically, but that some employed a grapheme strategy in responding to the task.

The important implication from these findings is that the most unambiguous estimate of a child's level of phonological awareness using the tapping procedure will be obtained when the test is comprised entirely of nondigraph pseudowords. An estimate of phonological awareness would still be obtained for those children who used their knowledge of grapheme-phoneme correspondence rules to generate a graphemic representation of the word and then tap out the number of graphemes, since in order to apply the correspondence rules to words not seen before, these children would first have to segment the words into their constituent phonemic elements. The results of Tunmer and Nesdale's study clearly indicate that performance on nondigraph pseudowords differentiates those children who have achieved an awareness of the phonological units of spoken language from those who have not, at least when taps or counters are used to represent phonemes.

Conclusions

What conclusions can we draw from the results of studies which have assessed young children's phonemic segmentation abilities? Perhaps the least controversial conclusion is that, regardless of the task and procedure used, there is an age-related trend in the acquisition of such abilities. The more difficult question concerns the level of phonological awareness we can expect to find in children at specific ages. As we have seen, estimates vary between reseachers who claim that children as young as 4 years can show a high level of phonemic segmentation ability [e.g., Fox and Routh, 1975] to those who suggest that children younger than 7 years do not possess segmentation skills [e.g., Bruce, 1964]. We have suggested that this range of opinion reflects results which undoubtedly differ as a result of both the task and procedure employed.

It is important to note, however, that if we exclude from our discussion the results of studies which employed procedures that require children to deal with non-phonemic segments (i.e. syllabic segments such as *buh* rather than /b/) the range narrows considerably. Moreover, the estimate of the age at which young children can display segmentation becomes more conservative. For example, the Tunmer and Nesdale [1982] results suggest that most children need to be 6 years of age or older before they display a reasonable level of segmentation ability. However, as we have observed, the few procedures used to date which provide a relatively unambiguous assessment of phonemic awareness appear to have increased the difficulty level of whatever segmentation task the child is required to do. In fact, as was demonstrated in the study by Calfee et al. [1973], both the task and procedures used can be of such a level of difficulty that even 6-year-olds cannot show a satisfactory level of performance. Given that our analysis of such tests suggests that the increase in difficulty level might well depend on the number of operations the child is required to undertake, one implication might be that we need to develop a test which reduces these demands. Nevertheless, we are still brought back to the conclusion that regardless of the level of segmentation ability we might want to attribute to a child, it will be both task- and procedure-dependent.

What are the implications of this conclusion? Do we simply select a hard or an easy or a moderately difficult test (i.e., task and procedure) for the purpose of assessment? Or do we administer a battery of tests encompassing the whole range of difficulty to each child? At this stage the answers are unclear. Perhaps the pragmatic answer lies in selecting a test on the basis of its functional significance. For example, since one of the major concerns is with reading acquisition, we should possibly look for a test which is a good predictor of reading readiness or acquisition when the effects of intelligence are held constant. While this implies a need for a systematic assessment of available as well as new tests, our review also suggests that whatever test we select, we need at least to control for training, motivation and feedback effects.

On the issue of what the development of phonemic segmentation abilities depends upon, little can be said at this stage. Does it reflect the development of other more general cognitive and/or linguistic abilities? Does it depend upon specific cognitive and/or linguistic experiences? Some evidence bearing on these questions was reported in a study by Hakes et al. [1980]. One hundred children between the ages of 4 and 8 were administered a conservation test (a commonly used measure of cognitive development) and three metalinguistic awareness tasks, one of which was the phonemic segmentation task used by Liberman et al. [1974]. The data provided evidence of a positive relationship between performance on the phonemic segmentation task and the conservation task, suggesting some relationship between phonological awareness and cognitive development.

There is also some evidence that segmentation ability is related to socio-economic status [Wallach et al., 1977] and IQ [Rosner and Simon, 1971, but see Tunmer and Fletcher, 1981]. However, since the latter variables are typically highly correlated and relatively broad-based variables, the significance of these relationships is presently unclear. In addition, Rosner and Simon [1971] reported that despite the relationship to IQ, phonemic segmentation skills independently contributed to reading achievement [see also Tunmer and Fletcher, 1981]. Clearly, while such findings intimate that some individual difference variables might well underlie or be related to phonemic segmentation, definitive statements on the issue must await further research.

What does seem to be evident at this point is that segmentation abilities are influenced by exposure to reading instruction [e.g., Bruce, 1964; Wallach and Wallach, 1976], some forms of which include a large component of phonological awareness training [Byrne, 1981]. Moreover, as Golinkoff [1978] points out, contrary to some researchers who assumed that phonemic segmentation skills could not be trained [e.g., Gleitman and Rozin, 1973], several researchers have now demonstrated that young children's segmentation skills can be improved through training programs [e.g., Rosner, 1971; Elkonin, 1973; Marsh and Mineo, 1977; Williams, 1980]. Since such training has also been shown to have a positive impact on reading achievement [e.g., Rosner, 1971], the implication for those interested in enhancing children's reading acquisition seems clear.

2.2 Word Awareness in Children

Judith A. Bowey and William E. Tunmer

To literate adults, the word appears as an obvious unit of language. However, to a large extent, this obviousness is more apparent than real, being the result of years of seeing words in print separated by spaces. This suggestion is well illustrated by the difficulties experienced even by linguists in attempting to define the term *word* [e.g., Lyons, 1968; Kramsky, 1969].

However, despite the fact that the word appears to resist both linguistic and acoustic definition and the possibility that word awareness may be enhanced by literacy, the ability to segment speech into word units does not appear to be dependent upon the prior ability to read. Sapir [1921] reported that he taught illiterate Indians to write their own languages using his phonetic system, and that they had no difficulty in determining the word units of their language. This pre-literate awareness of words as units of language would have been employed in the development of writing systems which use words as the units in print [Gelb, 1963]. Words may thus be viewed as "psychologically real" units of language, independent of literacy. Nevertheless, the difficulties involved in defining the word linguistically suggest that word awareness may not be easily acquired.

A fully developed word awareness would involve the following three components:

(1) awareness of the word as a unit of language.
(2) awareness of the word as an arbitrary phonological label (the word-referent distinction).
(3) comprehension of the metalinguistic term *word*.

There is no reason to suppose that these three components of word awareness emerge simultaneously. For instance, some workers have studied children's comprehension of the term *word* by asking them to define *word* [e.g., Berthoud-Papandropoulou, 1978; Papandropoulou and Sinclair, 1974]. However, the ability to define the term *word* involves an extremely high level of linguistic awareness: the problems experienced by linguists in developing criteria for segmenting speech sequences into words illustrate this point only too well. In this sense, full word awareness may be virtually unattainable.

It is obvious, then, that we must differentiate degrees of word awareness. Even within each of the three components of word awareness we must recognize varying degrees of attainment. For instance, it seems highly likely that most adults could provide a definition of *word* that would correctly differentiate a large proportion of words from other linguistic units, such as phonemes and phrases.

The rate of development of the three aspects of word awareness is likely to vary considerably. For example, the attainment of the "word unit" and the word-referent distinction is logically prior to the ability to comprehend or define the meaning of *word*. Even allowing for varying degrees of comprehension of the term *word*, this argument still holds, since a minimal understanding of the term *word* would involve the prior acquisition of the "word unit" concept. This argument follows Lenneberg's [1967] conclusion that the categorization of objects must precede the ability to name them. Objects must be categorized (correctly or incorrectly) before they can be named: "the abstractness underlying meanings in general....may best be understood by considering concept-formation, the primary cognitive process, and naming (as well as acquiring a name) the secondary cognitive process" [p.333]. Elsewhere, Nelson [1973] has found that children younger than 2 years old can "sort or group objects according to a consistent principle before they have adequate language to name the groups formed, to identify their basis of classification, or to understand classifying instructions" [p.28].

This argument concerning the logical priority of word concept attainment relative to comprehension of the term *word* is an extremely important one. It enables a crucial distinction to be made between the attainment of the concept of word and the understanding of the metalinguistic label *word*. Thus, a child might be able to segment a speech sequence into words, thereby demonstrating an awareness of the "word unit" concept, without knowing what is meant by the term *word*. A child might also use the term *word* without sharing the adult's notion of what this term means (see below).

The need to distinguish between, for instance, children's word unit concepts and their comprehension of the term *word* is clearly illustrated in a series of experiments carried out by Bowey et al. [in press]. They asked pre-school, first and second-grade children to distinguish words from sounds, and single words from two-word noun phrases in aural discrimination tasks. Children in an experimental group were introduced to these tasks by means of 12 practice trials incorporating both feedback and simple verbal explanations of the metalinguistic terms, *word* and *sound*. A control group was introduced to the tasks by means of analogous, rather than task-specific training on animal-nonanimal and fruit-nonfruit aural discrimination tasks. The control group was thus tested on comprehension of the term *word*, in addition to their concept of the word as a unit of language. The experimental group had been given an opportunity to attach the metalinguistic term *word* more securely to their existing word concept, and thus their performance in subsequent test trials on novel items was a purer index of their word unit concept. Not surprisingly, the experimental group performed at a substantially higher level than the control group on test trials.

The distinction between the attainment of the concept of word and the label *word* is critical in evaluating the literature on the development of word awareness. Many studies have incorporated the term *word* in experimental instructions without either attempting to define it or providing adequate preliminary training to ensure that the child could understand the task independently of his comprehension of the metalinguistic term *word*. These studies have presupposed an understanding of the label *word* in assessing logically prior aspects of word awareness. The conclusions that can be safely drawn from such studies are thereby highly restricted; they could well have substantially underestimated children's word awareness. Since these studies cannot be properly evaluated until we know what children understand by the term *word*, this review will commence with investigations concerned with this particular aspect of word awareness. Consideration of the development of children's awareness of the word as a unit of language and their awareness of the word as an arbitrary phonological label will then follow.

Comprehension of the Metalinguistic Term *Word*

The correct use of the term *word* has been observed in the speech of children as young as 1;9 [Bohn, 1914] and 3;4 [Slobin, 1978]. However, although the term may sometimes be used in a manner that is consistent with adult usage, it cannot necessarily be assumed that the child shares the adult's understanding of the term. A child may be aware of some aspects of the meaning of *word*, but confused about other aspects. Reid [1966], for instance, reported a 5-year-old as saying "words are made up of words" [p.61].

In some contexts, *word* may be correctly used by young children, but this does not guarantee correct usage or comprehension in other contexts. This suggestion is demonstrated by a close examination of Slobin's daughter's comprehension of *word* [Slobin, 1978]. At 2;10, Heida frequently asked for Czech equivalents of English words. She first asked only for nouns. Slobin then suggested "another kind of word, like *walk* or *eat*". Heida duly produced verbs, but soon asked for verb phrases. However, she could not play that game with adjectives. Thus, Heida's understanding of *word* was both overgeneralized (to include verb phrases) and undergeneralized (to exclude adjectives).

The conclusion that children's comprehension and correct usage of *word* in some contexts does not necessarily imply full comprehension is wholly consistent with recent work in language acquisition, where it has been noted that children both over and under-extend the meaning of newly acquired words [Clark and Clark, 1977]. Further evidence that children both over and under-extend the meaning of the term *word* comes from more experimentally oriented studies.

At about 5 or 6 years of age, children in most countries commence formal reading instruction, in which they are exposed to the technical vocabulary of reading. As a consequence, children's comprehension of the term *word* improves steadily. Francis [1973] asked children to "tell me a word -- any word you know". At 5;9, 22 of her 50 subjects could do so, and a further 13 provided either a name or a number, which also constitute words and, as Francis notes, are "within the area of the concept and indicated at least a partial understanding" [p.20]. The frequency of children giving words that were neither names nor numbers increased with age, reaching 46 by 7;3.

Despite this general improvement, children do not acquire a clear understanding of the term *word* for some time. Thus children of 5 to 7 years of age often only accept contentives (e.g., nouns, verbs, adjectives) as words [Papandropoulou and Sinclair, 1974]. Functors (e.g., articles and auxiliary verbs) are not accepted as words.

Berthoud-Papandropoulou [1978] and Papandropoulou and Sinclair [1974] asked children to define *word* by posing the questions, "What is a word, really?" and "How do you know whether something is a word?". Children aged 4 and 5 years gave "definitions" which suggested that words were not seen as having an autonomous existence, but somehow *were* the objects or actions themselves: "although words do not have an identity of their own, they are accepted as being words because the objects or actions they refer to exist" [Papandropoulou and Sinclair, 1974, p.244]. The authors also reported that it was not until 6;6 to 8;0 that children viewed words as units, as part of a sequence of more than one word carrying meaning ("bits of a story"). From 8 to 10 years of age, children provided definitions which strongly suggested the influence of formal education: the answers given by children were "remarkably uniform, [and] most children use exactly the same expressions to define words" [p.247]. "A word is something that means something, it's written with letters" (child 8;3).

This type of study involving children's ability to define the term *word* is open to serious criticisms. In particular, the developing understanding of the term *word* is confounded not only with the influence of formal education, which produces, in older children, formula-like answers, but also with the development of the ability to provide definitions. Even if young children have a fully developed word concept, it is unlikely that they would be able to define such an abstract term as *word*, since they cannot adequately define concrete common nouns such as *knife* or *nail* [Litowitz, 1977].

Downing [1969, 1970, 1971] reported a study testing 5-year-old children's understanding of the term *word* by means of an aural discrimination task, in which the child was asked to say "yes" whenever the stimulus was a word, and "no" whenever the stimulus was not a word. Non-word stimuli consisted of non-human noises, phonemes, phrases and sentences. None of Downing's 13 subjects responded "yes" only to words. Five children responded uniformly to all stimuli, with four of them classifying all as words. A further three classified all human speech (phonemes, words, phrases and sentences) as words, while the remaining five children grouped all meaningful human vocalizations (words, phrases and sentences) as words. These findings suggest that 5-year-old children have a poor understanding of the term *word*.

This result was confirmed by Downing and Oliver [1974] when they replicated the original study with an extended age range. They found that all children over-extended the meaning of *word*, although the degree of over-extension was less in the oldest children, aged from 6;6 to 8;0. The youngest children (4;5 to 5;5) tended to answer "yes" indiscriminately. None of the children in the study showed full mastery of the meaning of *word*. Even in the oldest group, phonemes and syllables were still confused with words, although by then phrases and sentences were usually classified as non-words. Similar findings regarding the child's limited understanding of the term *word* have been reported by Johns [1979] and Kingston et al. [1972].

It can be argued, however, that the aural discrimination tasks used by Downing [1971] were inherently difficult for young children. In those studies, children were required to respond "yes" to a target stimulus only when it was *both* a target and one target only, and "no" when a stimulus was either a nontarget or contained more than one target stimulus. In other words, a child was required to classify the stimulus along two dimensions simultaneously. Such a task involves multiple classification skills, which are not fully mastered until middle childhood [Brainerd, 1978; Inhelder and Piaget, 1964]. It is therefore possible that children's performance on Downing's discrimination tasks reflects their mastery of multiple classification, in addition to their comprehension of the metalinguistic term *word*. This argument holds despite the attempts made by both Downing and Oliver [1974] and Johns [1979] to screen children for their ability to carry out analogous training tasks, as it is possible that the training tasks were mastered on the basis of rote paired-associate learning [see Bowey et al., in press].

Bowey et al. modified the aural discrimination tasks, requiring pre-school, first and second-grade children to distinguish word from nonword stimuli along only one dimension at a time. The classification skills required in an unidimensional discrimination task are fully mastered by children of this age. School-aged children who were introduced to this task by means of an analogous task, rather than through task-specific practice items (see above), performed at a slightly higher level than that reported in previous studies, with 60% of first-grade, and 70% of second-grade children able to distinguish *both* words from sounds, and single words from noun phrases. Children understood the term *word* better in relation to the word-sound distinction than in relation to the word-phrase distinction.

The higher level of comprehension of *word* observed by Bowey et al. is more notable in view of the fact that their word stimuli consisted of words from a variety of grammatical form classes, with equal numbers of content and function words. The previous aural discrimination studies used as target words only nouns, which are probably recognized as words earlier than other types of word [Berthoud-Papandropoulou, 1978; Papandropoulou and Sinclair, 1974; Slobin, 1978]. Bowey et al. in fact found that pre-school and first-grade children were more likely to recognize as words contentives (e.g., nouns, verbs, adjectives) rather than functors (e.g., quantifiers, prepositions, conjunctions).

Despite the higher level of comprehension of *word* reported by Bowey et al. the evidence still suggests a gradual refinement of children's understanding of the term *word* with age. None of the ten pre-school children in the control group studied by Bowey et al. understood the term *word*. By the end of first and second-grade, 65% of children in the control groups comprehended it in relation to both the word-sound and word-phrase contrasts. This improvement, together with the marked advantage of children given 12 task-specific practice trials, in which they were helped to attach the term *word* to their existing word-concepts (see above), suggests that comprehension of the metalinguistic term *word* may be relatively easily taught [see Bowey et al., in press; Bowey, 1983].

A more immediate implication of the conclusion that young children do not have a clear understanding of the term *word* concerns the evaluation of research investigating other components of word awareness. Studies which require children to comprehend the term *word* in order to perform the experimental task are likely to seriously underestimate children's attainment of these other aspects of word awareness.

Awareness of the Word as a Unit of Language

Before a child can learn to use the label *word* correctly, he must be aware of the word as a unit of language, larger than phonemes and smaller than phrases, and syntactically distinct from bound morphemes. Research into the development of the "word unit" concept suggests generally that this insight may not be easily attained.

Several language acquisition studies have reported that children appear to play with language by trying out different sentence types, by repeating newly acquired words, by repeating sentence frames, substituting one word for another, and so on. Word substitution play has been observed by several researchers [Bohn, 1914; Snyder, 1914; Weir, 1962]. The following examples are typical:

(1) Train go on track. Car go on track. Wheel go on track. Little wheel go on track [Snyder, 1914, p.421].

(2) What color blanket? What color mop? What color glass? [Weir, 1962, p.109].

Weir [1962] noted that noun substitution play was most frequent, although pronoun, verb and noun modifier substitution play was also observed. This finding could be interpreted as suggesting that nouns are viewed as words earlier than are other form classes. Weir, however, attributed the differential frequencies of the various types of word substitution to the relative incidence of these word classes in the syntactic frames used by children.

Care must be taken in the interpretation of word substitution drill as language awareness. To distinguish this type of monologue from actual language use we must employ criteria such as "whether the child is directing his utterance to any addressee with any possible communicative intent" [Clark, 1978, p.29] and whether the monologue is serving "to accompany....action" [Piaget, 1959, p.15]. By these criteria, (2) may be viewed as a possible example of word substitution play, being soliloquized after the child had been left to sleep. However, (1) may well have been simply an accompaniment of action. For word substitution drill to be regarded as evidence of word awareness, such play must also be interpreted as "deliberate" rather than "spontaneous" [Hakes et al., 1980]. For example, the child who produced (2) may have been responding to mental images relating to the day's events rather than deliberately substituting words into a particular linguistic structure. Language play must at least satisfy the requirements that content must be "subordinate to the linguistic form" [Dale, 1976, p.155].

Experimental investigations of the "word unit" concept have required children to analyse spoken sequences into their constituent words. These studies have generally reported that young children experience great difficulty in segmenting speech. However, the extent to which segmentation ability was confused with knowledge of the spoken term *word* is not always clear, making it difficult to evaluate the findings of such studies. For example, Karpova [1966] required Russian children aged between 3 and 7 years to repeat sentences and respond to the questions, "How many words are there?" and "Which is the first... second... third... word?". The children experienced great difficulty in correctly answering these questions. A few of the older children were able to segment sentences into their component words, although errors were often made with prepositions and conjunctions. In addition, some children broke some words into syllables. The same general pattern of results was obtained in similar experiments carried out with French-speaking children [Berthoud-Papandropoulou, 1978; Papandropoulou and Sinclair, 1974] and English-speaking children [Hall, 1976]. Hall reported that children aged between 4;8 and 6 achieved 48.7% of the maximum possible score.

It should be noted that although precise details of the experimental procedures were not always supplied, these studies all appeared to use the label *word* in the experimental instructions, without providing any examples of how the task should be performed. They may therefore have confounded children's knowledge of the metalinguistic term *word* with their ability to segment speech into word units. In doing so, they may well have underestimated children's awareness of words as linguistic units.

Most other studies in the literature have attempted to overcome this problem by providing practice trials and/or corrective feedback when testing children's ability to segment speech into word units. Huttenlocher [1964], for instance, asked children aged between 4;6 and 5 years to either reverse or segment two-item sequences. Segmentation involved repeating the first item, and then waiting for the experimenter to tap before repeating the second. Full details of the instructions and procedure were not reported, but corrective feedback was provided. The segmentation task was easier than the reversal task, a finding which is not surprising in view of the additional mental operations involved in the latter task. Overall, the children correctly segmented over 80% of the item pairs consisting of letters or numbers, words of like form class (nouns, verbs or adjectives) and semantically anomalous sequences (e.g., *table goes*). Of the grammatical sequences, 61% were correctly segmented.

A commonly used paradigm involves asking children to listen to spoken sentences, and to represent each word with either a tap or a token (e.g., a small cube). Studies using this paradigm have reported that first-grade children experience considerable difficulty, generally scoring at about 50% [Evans et al., 1979; Kingston et al., 1972; McNinch, 1971, 1974]. Holden and MacGinitie [1972], using essentially the same procedure, and providing explicit demonstration of the task and corrective feedback on four practice trials, reported that kindergarten children performed much better on contentives than on functors. This finding was replicated by Ehri [1975] in a study of children aged from 4;4 to 7;8. Holden and MacGinitie [1972] also observed that awareness of some functors, such as *to* and *is* seemed partly dependent upon the surrounding context. They suggested that the rhythmic qualities of an utterance might influence children's segmentation performance. Ehri similarly commented that factors such as stress location and morphemic structure affected children's performance, with children making additional (erroneous) responses for bisyllabic words in which the second syllable was stressed and, in particular, for bisyllabic words made up of double unbound morphemes (e.g., *airplane*, *outside*).

Only two studies have reported that young children can successfully segment word strings into the component units. In both, special precautions were taken to ensure that children's performance was not confounded with their understanding of the spoken term word. Fox and Routh [1975] avoided the problem of using the term *word* in their experimental instructions by asking children from 3 to 7 years of age to repeat "just a little bit" of a spoken utterance. One example was demonstrated (*Peter jumps*). Children were then tested on sentences two to seven words long. If they responded with phrases, they were then led to segment each multiple word phrase further. Using this procedure, Fox and Routh reported that 4-year-old children could segment seven out of eight test sentences, while even 3-year-olds could segment five. However, these results do not necessarily indicate that children can segment speech into word units. Children performed at the same level in further segmenting words into syllables. There is no strong reason to suppose that children necessarily distinguished words from phrases or syllables. The experimenter continued asking for further segmentation until word units were produced, and then stopped. The results obtained were a direct function of the experimental procedure.

Tunmer and Bowey [1981] carried out a series of three experiments designed to investigate factors potentially affecting young children's ability to segment speech. They incorporated explicit demonstration and corrective feedback of practice items into their pre-test procedure to ensure that children's performance on the word tapping task was not confounded with knowledge of the term *word*. The most notable finding to emerge was that even 4-year-olds segmented 73% of two and three-word strings correctly, while 5-year-old first-graders performed with almost 90% accuracy. Six-year-olds achieved 98% accuracy. Segmentation performance was relatively unaffected by variables such as string length, syllabic length of words, and string meaningfulness. However, the children responded more accurately to strings composed entirely of nouns and adjectives than to those composed of verbs and quantifiers.

In these experiments only words that contained one stressed syllable each were used. It was therefore possible that children employed a syllabic stress strategy in producing their responses, tapping once for each stressed syllable. This would result in artifactually high scores, since the phonemic property of stress is perceptually more salient than the abstract concept of word. Tunmer et al. [in press] therefore conducted a fourth experiment to test the hypothesis that children are capable of using syllabic stress as a basis for responding to the word tapping task.

In addition to real word strings, children were presented with strings comprising pronounceable nonsense words (e.g., *loust namp denster*). By definition the children could not use the notion of "smallest *meaningful*, cohesive and permutable unit of language" as the basis for segmentation, since the units were meaningless. The results showed that the children performed about as well on the nonsense word strings as on the real word strings. It is therefore possible that the scores in the previous experiments were artifactually high, since some of the children may have resorted to a stress strategy. Consistent with this possibility is the work of Holden and MacGinitie [1972], who found that the children tested tended to segment sentences at their major stress points.

A fifth experiment was therefore designed by Tunmer et al. to investigate two aspects of the possible role of syllabic stress in children's segmentation performance. The first concerned the frequently observed phenomenon that young children often fail to respond to some words. This is most common in relation to unstressed functors like the words *the* and *a* [e.g., Ehri, 1975; Holden and MacGinitie, 1972]. The second aspect concerned the opposite kind of error, that of producing more than one tap for single words in a segmentation task. As noted earlier, Ehri [1975] found that morphemic structure was often associated with this kind of error. Children tended to make additional erroneous responses for bisyllabic words made up of double unbound morphemes, both of which were stressed (e.g., *outside, airplane*).

A difficulty that is encountered in manipulating morphemic structure and syllabic stress separately is the inherent confounding between stress and number of morphemes for most compound words (e.g., *cowboy* contains two stressed syllables and two morphemes). Tunmer et al. overcame this problem by embedding bisyllabic compound or noncompound words in meaningful two or three-word phrases beginning with either stressed or unstressed functors (e.g., *his toenail, her baby, a peanut, the doctor*). The scores of the children and a qualitative analysis of their response patterns suggested that the basis on which they segmented meaningful word strings underwent two major changes prior to the attainment of a mature word concept. Children aged from 4 to 5 years performed rather poorly on the task, responding to the word strings primarily on the basis of such phonemic factors as phrase and syllable stress. An intermediate strategy corresponded to that commonly reported in the literature, in which function words are not responded to as words. Somewhat older children adopted a different strategy, tapping once for each unbound morpheme in the phrase. Both the content word and the unbound morpheme strategies represent relatively well-developed word concepts, in that syntactic distributional factors form the basis for responding and that these strategies are relatively accurate in segmenting normal discourse into constituent word units. By about 7 years of age, most children had begun to abandon these strategies in favour of an adult-like word unit concept based upon more accurate distributional properties. Compound words and function words were more reliably seen as words. Although not complete by age 7, the development of an accurate word concept, as revealed by the conservative requirement that children correctly segment noun phrases containing compound words was well under way.

While not as high as those obtained in the first four experiments, children's scores in the fifth experiment were still slightly higher than those reported elsewhere in the literature. It appears likely that, apart from frequently confounding segmentation ability with knowledge of the metalinguistic term *word*, previous researchers may also have exceeded children's memory capacity. Although their subjects could remember the sentences to be segmented, the strain on memory could well have been sufficient to disrupt segmentation performance. The task of repeating sentence-long strings while simultaneously identifying word boundaries may be too complex for young children [Lundberg, 1978].

Awareness of the Word as an Arbitrary Phonological Label

Understanding that the phonological realization of a word is essentially arbitrary, that is, that words are phonological labels attached to their referents by convention, appears to emerge somewhat later than the word unit concept.

This aspect of word awareness was first studied by Piaget [1929], who asked children aged between 5 and 11 years a number of questions concerning the origins of names and the potential interchangeability of object names. Although children of 5 and 6 years could distinguish the name from the object named, they could only conceive of the name as coming from the object. The name seemed to emanate from the object, as an invisible quality that is an essential part of the object. Thus the object did not exist prior to the name, and we came to know the names of objects simply by looking at these objects. Names for things could not be interchanged. Similar findings were reported by Vygotsky [1962], who noted that pre-school children "explain the names of objects by their attributes" [p.129]. An animal is called *cow* because it has horns. A cow cannot be called *ink* and ink *cow* "because ink is used for writing, and the cow gives milk" [p.129]. Piaget reported that 7- and 8-year-old children alleged that names were invented by the makers of the object, by God, or by the first men. Although the name might be given by men, it was given with the object, so that it is "consubstantial" with the object. Names were frequently dissociated from the object, but were not yet seen as conventional labels. The name contained the idea of the object, so that names for objects could not be interchanged. By 9 and 10 years of age, children viewed names as given by men of no fixed identity, being passed down from generation to generation. The name was no longer identified with the idea of creation: objects existed before they had names. Names were seen to be conventional -- they could have been different. Some children, however, did not regard the name as arbitrary: the name somehow contained the idea of the object so that it "fits" well. It was only by 10 or 11 years that children viewed the name in itself as nothing but a sign.

More recently, experimental studies have suggested that children's ability to differentiate words from referents may originally have been underestimated. In particular, the wording of questions is critical. Ianco-Worrall [1972] found that only 8% of 4, 5 and 6-year-olds answered "yes" to questions such as "Could you call a dog *cow* and a cow *dog*?". However, when the question was changed to "Suppose you were making up names for things, could you then call a cow *dog* and a dog *cow*?", 16% agreed that this could be done. This figure reached 38% in 7-, 8- and 9-year-olds. Similarly, Osherson and Markman [1975] reported that all children in their sample, which included children from grades 1, 2, 3, 6 and 10, "acquiesced with minimal coaching" to the question "Suppose everyone in the world got together and decided that from now on we will call the sun *the moon* and we will call the moon *the sun*. All we are going to do is change the names. Could we do that if we wanted to?". Virtually all subjects further responded that the thing we see in the sky at night was called *the sun*.

There is some evidence that children's performance on word-referent tasks is also dependent upon the processing demands of the task. Osherson and Markman found that although children agreed to substitute *sun* for *moon* and described the thing seen in the night sky as *the sun*, they could not describe this new night sky as dark. However, the task of describing the new night sky involves considerable processing strain. It involves not only the dissociation of the word *moon* from its referent, but also the reversal of overlearned associations between the phonological realizations of both *sun* and *moon* and their semantic features. The difficulty of this task can be illustrated by asking an adult to tell a story involving the interaction of a cat and a dog, substituting the word *cat* for *dog* and *dog* for *cat*. Such a task involves much more than an acknowledgement that names are essentially arbitrary.

The argument that children's performance on word-referent tasks varies with the processing demands of the task is supported by a study carried out by Feldman and Shen [1971]. They found that 5-year-old monolingual children learned to call an object by a nonsense label (e.g., *wug*) 93% of the time. However, they could substitute common names (e.g., call a cup *plate*) only 31% of the time. In other words, in a task requiring only the implicit acknowledgement that names are arbitrary, children performed at a higher level than in a task requiring the exchange of two sets of labels and meanings. When task demands were increased by requiring children to use these new referent labels in simple relational sentences (e.g., "The cup is on the plate"), success was much lower, with nonsense labels being used as poorly (17%) as substituted common nouns (13%).

Piaget [1929] also investigated children's ability to distinguish word and referent by posing the question, "Can a word have strength?" and asking children to give a "strong word". Children aged between 7 and 8 years made no distinction between word and referent, and failed to understand that the question constituted a problem. At 9 years of age, children had reached a "paradoxical stage" where, although they understood the problem inherent in the question, the word-referent distinction was not sufficiently developed to save the child from falling into the trap, and supplying a "strong word". By

10 and 11 years, children were fully aware of the misleading nature of the question.

Papandropoulou and Sinclair [1974] and Berthoud-Papandropoulou [1978] extended this work by asking children to give a "long word", a "short word" and a "difficult word". Young children could not distinguish word from referent, supplying the name of a long or large object when asked to give a long word. Similarly, they would describe a difficult or unpleasant action or object for the "difficult word" task. Children aged 5, 6 and 7 years supplied a number of propositions when asked to give a long word, and a single clause or sentence for a short word. They claimed that they could not supply a difficult word, because that itself was difficult, but would name an easy word like *cat* because it was "easy to say". From 6;6 to 8 years, most used adult criteria such as the number of syllables or letters to distinguish long and short words. By 8 years, children used adult criteria to provide difficult words, giving multisyllabic recondite words.

However, since both Piaget and Papandropoulou and Sinclair phrased these questions using the term *word*, which, as we have already seen, is not clearly understood by young children, it is probable that children's ability to distinguish word from referent was confounded with their failure to comprehend the term *word*.

A similar problem is encountered in the work of Osherson and Markman [1975], who also studied children's awareness of the non-physical reality of words. They asked children a series of four questions of the form, "Is the word *book* made of paper?". Children were required to answer all four correctly to be viewed as passing this test. Of the 24 children tested from the first three grades of school, only two passed. Twelve of the 15 sixth and tenth-grade children passed. However, only eight of the 11 adults tested passed. Furthermore, two additional questions were not scored in the study "*since adult subjects reported confusion as to whether the questions should be interpreted literally or metaphorically*" [p.221, emphasis added]. (These additional questions were "Is the word *fire* hot?" and "Is the word *ice* a cold word?"). The finding that not all adults passed this test and the discarding of the two additional questions, together suggest that the word-referent test used by Osherson and Markman may well be far too difficult for children. They, too, are likely to experience confusion concerning the literal or metaphorical interpretation of the question.

This suggestion is made more plausible in view of the evidence that children do not fully understand the term *word*. Donaldson [1978] has suggested that classical Piagetian tasks may well underestimate children's ability in that children may not interpret the question as the adult intended it. It should be noted in this context that young children will answer questions that they do not, or even cannot, understand. Hughes and Grieve [1980] found that children will provide answers to conceptually ill-formed or "bizarre" questions, usually by importing additional context. They conclude that:

> the fact that the child will attempt to locate an answer to whatever question he is presented with has significant implications for what we think is happening when we attempt to gauge the young child's cognitive/linguistic abilities by means of the question and answer process. Psychologists...can no longer treat the child as merely a passive recipient of questions and instructions, but must instead start to view the child as someone who is actively trying to make sense of the situation he is in -- however bizarre it may seem. [p.161]

It is therefore not surprising that children answer questions such as "Is the word *book* made from paper?" even if these questions are not clearly understood. However, great caution needs to be exercised in the interpretation of these answers.

The view that children's poor performance on this type of task is at least partially attributable to their interpretation of the question is supported by a further study by Markman [1976]. She found that children responded much better to less ambiguous questions such as "What has wheels, the word *car* or a car?" than to standard questions such as "Does the word *car* have wheels?". Markman [1976] noted that, in word-referent studies, "it seems as though children are consistently answering questions about empirical objects and ignoring the fact that the actual questions are about words" [p.743].

This is hardly surprising when it is recalled that children lack a precise understanding of the term *word*. Markman [1976], however, attempted to account for children's failure to interpret such questions appropriately primarily in terms of the "transparency" of language, whereby one looks through language to communicate about reality. Furthermore, the character of words is ephemeral and intangible when compared to the permanency and tangibility of physical objects.

Markman [1976, p.744] therefore hypothesized that "the more salient empirical properties of the referents" dominated children's thought. This would account for their failure to distinguish word from referent, and yields the prediction that children would be less likely to make the "converse" error of attributing properties of words to their referents. This hypothesis was tested by comparing children's performance on standard questions (e.g., "Does the word *car* have wheels?") with "converse" questions (e.g., "Does a car start with *kuh*?"). However, the hypothesis was supported only by data from first-grade children, and even then converse questions were correctly answered only 26% of the time. Second-grade children actually performed more poorly on the converse than on the standard questions. Markman attributed this result to second-grade children's greater phonological awareness, which allowed them to attend to sounds of words rather than meanings. However, if this was true in the converse task, it should also have been true in the standard task, so that they should have performed considerably better than the 45% level that was, in fact, observed.

The fact remains that Markman's hypothesis was supported only in relation to first-grade children and cannot account for the poor performance by second-graders on standard word-referent questions. The most plausible explanation for children's poor performance on these questions appears to be that they simply do not comprehend the question, because they do not have a clear understanding of the term *word*.

The possible reinterpretations of most of the word-referent studies suggest that they may have considerably under-estimated children's ability to distinguish word from referent. In particular, many of the tasks reviewed have employed questions which children find difficult to interpret. Thus, it seems reasonable to expect that, given more suitable tasks, children might demonstrate an ability to view words as separate from the objects they label.

A possible alternative to the traditional word-referent tasks is provided by lexical ambiguity tasks, which test children's awareness that at least two referents share a common phonological realization (e.g., "The man picked up the pipe": to smoke, or to mend the plumbing). Studies of children's awareness of lexical ambiguity may be interpreted as indicating their ability to separate word from referent, since detection of lexical ambiguity involves the knowledge that at least two referents share a common phonological realization. This in itself requires the ability to dissociate meaning from the word, for unless this can be done, the alternative meaning can never be perceived.

Recent experimental evidence has shown that quite young children are indeed aware of lexical ambiguity, although the rate of detection varies greatly from study to study [see Brodzinsky et al., 1977; Goldstein, 1976; Kessel, 1970; Shultz and Pilon, 1973]. Since these studies have used similar tasks, in which children were required either to paraphrase lexically ambiguous sentences or to indicate pictures representing those sentences, the differences in detection rates can probably be attributed to differential meaning bias [see Hirsh-Pasek et al., 1978] or, possibly, to the differential ease of pictorially representing some meanings relative to others [Kessel,

1970]. Although the detection rates obtained in the various studies cannot be considered particularly valid, these studies do indicate that even kindergarten children are aware that lexical ambiguity exists [Kessel, 1970]. All studies reviewed reported a growing awareness of lexical ambiguity in the first few years of schooling.

It is probably no coincidence that appreciation of linguistic ambiguity in the form of riddle-play develops rapidly in the first few years of school, reaching a peak in third grade [Sutton-Smith, 1976]. Sutton-Smith asked children in grades one to eight to tell him riddles. Pre-riddles, or non-riddles, which share the question and answer form of true riddles, but which have arbitrary, idiosyncratic answers, rather than the systematic reclassification answers found in true riddles, constituted about a third of the responses obtained from first and second-grade children. Bowes [1979] reported a sharp decline in the frequency of non-riddles from first to fourth grade. These findings suggest an increasing awareness of riddles as speech play in the early years of school.

Bowes also found that most riddles provided by children involved lexical ambiguity. Of the riddles given by first-grade children, 52% played upon lexical ambiguity, with this figure reaching a peak of 63% in third grade. Brodzinsky [1977] found that fourth-grade children averaged 90% in their spontaneous comprehension of riddles involving lexical ambiguity. A similar level (for prompted comprehension) was reported by Hirsh-Pasek et al. [1978], with near-maximum performance being observed in fifth and sixth-grade children. Hirsh-Pasek et al. took care to exclude from consideration any jokes in which the child did not understand both meanings of the ambiguous words (as determined after the comprehension test). Using this procedure, they found that even first-grade children understood approximately 72% of jokes involving lexical ambiguity.

Riddle studies thus confirm the suggestion that children are aware of and, in fact, delight in lexical ambiguity, given an appropriate context. It should be noted, however, that riddle studies do not provide particularly pure measures of word awareness, probably underestimating the ability of younger children. Familiarity with both riddle-telling behaviour and the riddles themselves probably increases with age [Fowles and Glanz, 1977]. Similarly, appreciation of lexical ambiguity is related to vocabulary development, which also is age-dependent. Furthermore, the cognitive demands of the riddle comprehension task are considerable:

> An incongruity is created between the expected answer and the actual answer. The listener must then go back, locate the source of this incongruity...and develop a new rule relating question to answer. [Fowles and Glanz, 1977, p.437]

Although riddle studies cannot be regarded as pure measures of linguistic ambiguity detection, they do serve as a useful supplement to other paradigms in indicating that young children are aware that lexical ambiguity exists. Furthermore, despite the cognitive complexity of the riddle comprehension task, it appears that linguistic ambiguity is detected most readily in the riddle context. Hirsh-Pasek et al. [1978] point out that, in riddles, contextual cues are used to bias the listener in one direction in the "set-up" and then in the other direction for the "punch-line", facilitating the detection of linguistic ambiguity.

From this rather extensive treatment of the word-referent literature, it is concluded that quite young children appear capable of making the distinction between word and referent and of dissociating the two. The classical word-referent paradigms, based on the work of Piaget [1929] and Vygotsky [1962], seem to underestimate children's abilities --it is uncertain whether children interpret such tasks as adult experimenters intend them. It has been seen that more recent work, particularly that involving the detection of lexical ambiguity, suggests that children can indeed separate word from referent. However, there is a need for further work in this area. Considerable effort needs to be directed towards the development of new and unambiguous tasks. The extension of the linguistic ambiguity paradigm to even younger children may be instructive, although the lower level of vocabulary development in these children might render this task difficult.

Conclusions

In this chapter, word awareness has been viewed as consisting of at least three aspects: comprehension of the metalinguistic term *word*, the attainment of the "word unit" concept and awareness of the word as a "phonological label". It was argued that these three aspects of word awareness may develop relatively independently, emerging at different times.

Failure to acknowledge this possibility has led many researchers to experimentally confound one aspect of word awareness with others. In particular, many have presupposed adult-like comprehension of the term *word* in studying logically prior aspects of word awareness. Since available evidence suggests that even 8-year-old children may lack an accurate understanding of the term *word*, procedures which presuppose this understanding in testing the "word-unit" or the "phonological label" concepts are likely to underestimate children's abilities.

A second serious problem with much of the previous research concerns the relatively impure nature of the tasks used to assess word awareness. For instance, the development of the concept of the word as a unit of language has almost invariably been tested using sentence-long materials. The task of simultaneously recalling sentences and segmenting them into their component words is likely to be too difficult for younger children, regardless of their awareness of words as units of language. Similarly, studies of children's ability to distinguish between word and referent have frequently tested this by asking children questions which even adult subjects have sometimes regarded as ambiguous. When the ambiguity of the questions is lessened, children's performance on these tasks increases.

Future research should focus on the development of experimental tasks that avoid the problems of confounding the various components of word awareness and of confusing the development of word awareness with other aspects of cognitive development. Only then will an accurate assessment of the development of word awareness in children be possible.

2.3 Syntactic Awareness in Children

William E. Tunmer and Robert Grieve

The focus of the present chapter is on the child's emerging ability to reflect upon the internal grammatical structure of sentences. This we have called syntactic awareness. The chapter is divided into three sections. The first section discusses the early work on the development of children's awareness of the syntactic and semantic properties of sentences; the second examines a number of methodological and conceptual issues raised in recent articles on form awareness in children; and the third briefly considers two closely related research topics--children's awareness of structural synonymy (the property that two superficially different sentences share the same underlying structural representation), and children's awareness of structural ambiguity (the property that a given surface string has associated with it two or more underlying structural representations).

The Early Work

Adult speakers of a language not only possess the ability to produce and understand utterances, but in addition they can judge whether or not sentences are grammatically well-formed and semantically coherent [Gleitman and Gleitman, 1970]. While adults are not always in complete agreement about *which* sentences are acceptable in a language (indeed, they are often unable to give adequate reasons for judging particular sentences to be unacceptable), it is clear that the *ability* to make acceptability judgements is a part of the adult's linguistic competence. An important source of data that the linguist uses in formulating grammatical theories is the set of sentences that the adult judges, upon reflection, to be well-formed. Comparable data from child informants would be useful to the developmental psycholinguist attempting to describe the young child's linguistic competence, for it would afford a means of determining whether or not a particular syntactic rule should be included in the child's grammar. However, acceptability judgements are difficult to obtain from young children, as anyone who has tried to elicit such judgements knows. When Brown and Bellugi [1964] attempted to elicit such a judgement from a 2 -year-old, the result was other than anticipated:

> Another week we noticed that Adam would sometimes pluralize nouns when they should have been pluralized and sometimes would not. We wondered if he could make grammatical judgments about the plural, if he could distinguish a correct form from an incorrect form. "Adam", we asked, "which is right, 'two shoes' or 'two shoe'?" His answer on that occasion, produced with explosive enthusiasm, was "Pop goes the weasel!" [p.135]

It is certainly difficult to get young children to give stable judgements about their own language. The question, of course, is whether young children are unable, or merely unwilling, to provide such judgements. One possibility that can be considered is that while young children do possess the ability to make judgements of well-formedness, they simply do not understand the question that is put to them. If this is so, an appropriate procedure for eliciting acceptability judgements from young children needs to be found.

One procedure, developed by Gleitman et al. [1972], involved the use of role-playing to convey the difficult instructions. Each of three girls, all about 2-years-old, was asked to judge whether sentences sounded "good" or "silly" in the context of a game with the child's mother. Well-formed and telegraphic imperatives were used, together with their inverted forms (e.g., "Bring me the ball", "Bring ball", "Ball me the bring", and "Ball bring"). All three children judged the well-formed imperatives "good" more often than chance (92%, 80% and 80%), and were more likely to say that the inverted imperatives were "silly" than to say that the normal ones were. But even the inverted imperatives were more likely to be judged "good" than to be judged "silly" (75%, 50%, 58%). The telegraphic imperatives presented in normal order were also judged "good" more often than chance (100%, 82%, 58%), as were the inverted telegraphic imperatives, though not to as great an extent (58%, 58%, 58%). These results suggest that children under 3 years of age are not wholly incapable of distinguishing well-formed sentences from deviant ones. However, the children's criteria for accepting sentences appeared to be more lax than those of adults, in that they accepted many sentences that adults would not.

This gives rise to the question of the basis on which such judgements are being made. In spontaneous speech, children of this age are clearly using word order for the purpose of conveying semantic relations. Further, this word order is used by adults in working out what children mean [Brown, 1973]. However, the fact that children use word order in producing sentences does not mean that they actually contemplate the constraints on arrangements of words and phrases.

In addition to judging the acceptability of sentences, the children in the Gleitman et al. study were asked to "fix up" the silly ones. De Villiers and de Villiers [1974] argue that acceptability judgements themselves tell us very little about the basis of such judgements, suggesting that explanations of corrections are far more revealing. They note that of the few corrections obtained by Gleitman et al., the majority were semantic changes (e.g., "Put on a coat" to "Put me in a coat"). While a few corrections changed both semantics and word order (e.g., "Box the open" to "Get in the box"), only three of the 19 corrections changed word order alone. This led de Villiers and de Villiers [1972] to question the conclusion drawn by Gleitman et al. that it is possible to obtain adult-like judgements of grammatical well-formedness from 2-year-old children.

Using a modification of the Gleitman et al. procedure, de Villiers and de Villiers [1972] conducted an experiment designed to test the hypothesis that semantic factors play a more important role than syntactic factors in children's acceptability judgements. The task required children to judge and correct utterances produced by a puppet who sometimes "said things all the wrong way round". The puppet spoke well-formed and anomalous imperatives, as well as correct-order and reverse-order ones (e.g., "Throw the stone", "Throw the sky", "Brush your teeth", and "Teeth your brush"). Eight children were tested, ranging in age from 28 to 45 months.

Nearly all the children judged the anomalous imperatives to be "wrong" more often than they did the well-formed counterparts. But only the more linguistically mature children, as measured by their mean length of utterance (MLU), judged the reverse-order imperatives to be "wrong" more often than their correct-order counterparts. The least advanced children appeared to ignore the faulty grammar of the reversed imperatives, basing their responses on semantic considerations instead. For example, when presented with the word string, "Teeth your brush", one child immediately performed the action of brushing his teeth, and judged the string to be "O.K.". Slightly more advanced children began to reject some of the reversed imperatives, but when asked what was "the right way to say it", they still seemed to be guided more by considerations of meaning (e.g., "House a build" was corrected to "Live in a house"). On the basis of these findings, the authors concluded that semantic factors predominated in determining the children's judgements, and that the young children could not make correct judgements of syntactic acceptability.

The pattern of results obtained by de Villiers and de Villiers suggests that young children make judgements on a different basis from that of adults. If young children can make sense of a sentence, they will accept it;

if they cannot, they will reject it. In support of this interpretation are several studies of young children's comprehension performance which have found that young children are relatively insensitive to the order of words in utterances [Bever, 1970; de Villiers and de Villiers, 1973; Maratsos, 1974]. Bever [1970] presented children aged between 2 and 6 years with reversible active and passive sentences (e.g., "The alligator chases the tiger" and "The tiger is chased by the alligator"), asking the children to act out the sentences with toys. The performance of the youngest children was near chance for both kinds of sentences. For the active sentences, performance steadily improved with increasing age, so that by age 5 years the children's performance was nearly perfect. For the passives, however, performance first became worse than chance, and only later improved to a better-than-chance level. Bever concluded that children between 3 and 4 years of age were over-generalizing the word-order rule of the more frequent active sentences, and that the youngest children had not yet adopted the strategy of using word-order for comprehending utterances. This would suggest that 2-year-olds rely primarily on the meanings of the words in a sentence, fitting them together in some way that makes sense.

The performance of young children on the acceptability tasks used in the de Villiers and de Villiers [1972] and Gleitman et al. [1972] studies can be explained in terms of the comprehension strategies available to children of this age. Children who do not yet use word-order information in comprehension should "understand" both well-formed and reverse-order imperatives. But since there is no arrangement of the words of anomalous imperatives that makes sense, these will not be understood. Later, when the strategy of using word order in comprehending sentences is more developed, children should also not be able to understand reverse-order imperatives. Hence, the fact that the de Villiers' younger subjects rejected only the anomalous imperatives, while their more mature subjects rejected both the anomalous and reverse-order imperatives, follows directly from the nature of young children's developing comprehension strategies. The data suggest that young children accept sentences that they think they understand, while rejecting those they find incomprehensible.

Recent Studies

Several more recent studies have examined acceptability judgements in children, but have tended to focus on children typically 5 years of age and older [Bohannon, 1975, 1976; Howe and Hillman, 1973; James and Miller, 1973; Scholl and Ryan, 1975, 1980]. Scholl and Ryan [1975] were sharply critical of the experimental procedures used in the Gleitman et al. and the de Villiers and de Villiers studies, especially the use of "good" and "silly" in the former study, and "right" and "wrong" in the latter. The de Villiers had children judge whether an imperative spoken by a puppet was "right" or "wrong" instead of "good" or "silly" because, they argued, "good" and "silly" have semantic connotations. However, Scholl and Ryan argued that the use of "right" and "wrong" also might have biased the children's responses by encouraging them to think about the sentences in terms of whether or not what they asserted was a (morally) wrong thing to do.

Another methodological error appearing in both the de Villiers and de Villiers [1972] and Gleitman et al. [1972] studies, according to Scholl and Ryan, was the differential reinforcement of children's responses. The children in these studies were asked to correct any stimulus that was judged "silly" or "wrong", a potentially aversive situation which may have biased children's responses in the direction of judging sentences as "good" or "right". In an attempt to avoid the problems of earlier studies, Scholl and Ryan used a nonverbal, forced-choice procedure in which children were asked to point to the adult or the child in a photograph, depending upon which person was judged to have spoken the stimulus sentence. Further, to avoid biasing children's responses, they were not asked to provide corrections after their classifications of stimulus items. Using negative and interrogative sentences of varying grammatical complexity, Scholl and Ryan found that 5- and 7-year-old children did show some ability to discriminate well-formed sentences (e.g., "We can not go home", "What can the cow say?") from "primitive" (i.e., deviant) ones (e.g., "Not we go home", "What the cow say?"). They also found that performance improved with age on the negative sentences, but not on the wh-word questions. Overall, however, the performance of both age groups was rather poor.

In a later study Scholl and Ryan [1980] revised their task instructions and practice sentences to decrease the chances of misleading the children about the nature of the task. In their original study they had used practice sentences which may have encouraged judgements based on the content of the sentences, rather than on their form. In particular, interviews with the children after presentation of the test stimuli indicated that many assigned the questions to the child in the photograph because they believed that only small children would ask questions. With their revised procedure, Scholl and Ryan found that both age groups of children performed at much higher levels, and that the discriminations of the 7-year-olds were significantly better than the 5-year-olds on both the negative and interrogative sentences.

Bohannon [1976] used a somewhat similar procedure to investigate children's ability to discriminate between normal and scrambled sentences. Children from 5 to 7 years of age were presented with pictures of two adults, and asked to identify normal sentences as having been spoken by one of the two adults, and scrambled sentences by the other. Similar to Scholl and Ryan [1980], Bohannon found that performance on the task increased significantly with age. However, the children at all age levels did not perform as well as those in the Scholl and Ryan study. On the basis of a criterion of four errors or less (out of 24 trials), only 22% of kindergarteners were "discriminators", as compared to 58% of first graders and 78% of second graders.

The task used by Scholl and Ryan seems to be a more natural and less demanding task for children, and the difference in the results obtained in the two studies may well be attributable to this. As several studies of children's communication abilities in relatively unstructured situations have shown, children as young as 4 years are capable of adjusting their messages on the basis of the audience they are addressing [Gleason, 1973; Gelman and Shatz, 1977; Shatz and Gelman, 1973; Sachs and Devin, 1976]. Shatz and Gelman [1973] for example, found that 4-year-olds used shorter, simpler sentences when speaking to 2-year-olds than to other 4-year-olds or adults. It seems likely that the task used by Scholl and Ryan, which required children to attribute well-formed sentences to an adult and primitive (childlike) sentences to a 3-year-old, would have drawn on this ability, whereas the task used by Bohannon would not. This would explain the superior performance among the children of the Scholl and Ryan [1980] study.

Acceptability tasks involving semantic restrictions have also been studied [Howe and Hillman, 1973; James and Miller, 1973]. James and Miller presented two groups of children, aged 4;8-5;3 and 6;8-7;3, with sentences that were either normal or contained a single selection restriction violation of the features +animate or +human. The two types of restriction violation involved either subject-verb or adjective-noun constituents (e.g., "The large rock walked down the hill", "The happy pencil rolled off the desk", "The big spider skated across the room", and "The furry girl smiled at the man"). The children were asked to identify the anomalous and meaningful sentences as either "silly" or "okay". The results indicated that both 5- and 7-year-old children were capable of distinguishing between meaningful and anomalous sentences involving the +animate or +human selection restrictions.

Howe and Hillman [1973] also investigated the acquisition of the +animate selection restriction in children. Two kinds were examined, those involving the sentential subject and those involving the sentential object (e.g., "The story believed the teacher" and "The dog frightened the car"). Howe and Hillman began by presenting the children, aged 4 through 9 years, with grossly anomalous sentences and asking them to describe what was wrong with them. Having elicited the children's own terms, including "bad", "silly", "wrong", "stupid", "make believe" and "doesn't make sense", they then asked the children to evaluate pairs of sentences, and for each pair they were asked to decide which of the two sentences was unacceptable (i.e., "silly", "dumb", etc.). It was found that the acquisition of the two kinds of +animate selection restrictions continued well into primary school, with the animate subject restriction being acquired earlier than the animate object restriction.

These findings would appear to conflict with those obtained in the James and Miller [1973] study, where no developmental difference was observed between the 5- and 7-year-olds in the identification of sentences containing violations of the +animate feature. However, the task used by Howe and Hillman [1973] was clearly more demanding, since children were required to evaluate *pairs* of sentences, selecting the one they judged to be unacceptable. It is possible that such a task exceeded the memory capacities of the younger children, thus accounting for their poorer performance. In their pilot work, Howe and Hillman discovered that the task used for children aged over 5 years was not appropriate for preschoolers (4-year-olds):

> Many children at preschool level seemed to respond solely on the basis of primacy or recency, always selecting the first or second sentence as unacceptable. Hence, a slightly different task was used with nursery school children. Rather than using both members of each pair, only one version of the pair was read. [p.135]

Given that the task was clearly too demanding for the 4-year-olds, it might well be assumed that many 5- and possibly some 6-year-olds also encountered difficulty. Moreover, if the strain on memory were indeed disrupting performance, it would explain the poorer scores on the sentences involving the animate object restriction violation, since these occur *later* in the sentence than the animate subject violation.

An important issue regarding the acquisition of semantic constraints concerns the specific character of the restrictions. As Howe and Hillman point out, the restrictions may be as abstract as they are for adults (e.g., "The rock saw Billy" is unacceptable because *rock* is inanimate). Alternatively, children may be making the same judgement, but on a more concrete basis, with personal experience or "knowledge of the world" playing a greater role (e.g., the sentence, "The rock saw Billy", is rejected because rocks do not have eyes). A recent study by Carr [1979] supports the latter view, which suggests that many investigators may have incorrectly assumed that questions of the form, "Does that sentence sound O.K.?" tap metalinguistic awareness rather than the child's ability to verify sentences in terms of his past experience.

Carr [1979] conducted a longitudinal study of children between the ages of 2 and 5 years to explore the developmental changes in their judgements about the acceptability of anomalous and non-anomalous sentences involving the +animate selection restriction. Carr concluded that the pattern of results she obtained could only be explained in terms of an experience-based verification strategy in which the child relates the meaning of a sentence to his experience and decides whether its content is verified. These findings are consistent with the work of Donaldson and McGarrigle [1974], who argue that when very young children assign truth values to statements, they are more constrained by nonlinguistic factors than lexical or syntactic rules. That is, young children seem to be more experientially oriented rather than concerned with aspects of the linguistic or logical structure of the material in question.

These conclusions are supported by a study by Hakes et al. [1980] in which judgements on a wide variety of grammatical and ungrammatical sentences were elicited from children between the ages of 4 and 8 years. For both groups of sentences, the mean percentage of correct responses increased fairly consistently with age. At all ages, performance was better on the grammatical than on the ungrammatical sentences. For the ungrammatical sentences, the change with increasing age represented a tendency to say that such sentences were not acceptable. This increase was interpreted as reflecting the children's increasing knowledge of the rules of the language. The more rules or constraints the children know, the fewer ungrammatical strings of words they will accept as being consistent with those rules.

This explanation of the improvement in the children's performance on the ungrammatical sentences, however, cannot account for their improvement on the grammatical ones, since the latter do not violate any rules. That is, knowledge or lack of knowledge of the constraints on sentences would not result in the rejection of sentences not violating any constraints. To understand why so many 4- and 5-year-old children rejected sentences that older children and adults found perfectly acceptable, Hakes et al. examined the explanations the children gave for judging grammatical sentences to be unacceptable. They found that the reasons given were typically "content oriented", in the sense that they involved an evaluation of the assertion made by a sentence rather than of the sentence itself. For example, a 4-year-old rejected the sentence "Yesterday daddy painted the fence" because "Daddies don't paint fences, they paint walls". Other explanations suggested that children were evaluating sentences in terms of the potential consequences which the actions described by the sentence might have (e.g., "The big rock was in the middle of the road" was judged unacceptable because "A car might run over it and get a flat tyre"). Hakes et al. concluded that the age-related increase in correct judgements for the grammatical sentences can be explained by the tendency of the younger children to judge sentences on the basis of what they asserted rather than the linguistic manner in which they did so.

These findings, and those reported by de Villiers and de Villiers [1972] which were discussed earlier, suggest that the basis on which children judge acceptability undergoes two changes before children reach adulthood. Children from 2 to 3 years of age appear to judge sentences in terms of whether or not they are understood, accepting sentences they think they understand, while rejecting those they find incomprehensible. Somewhat older children, aged 4 to 5 years, adopt a content criterion, rejecting many sentences that they understand but which say things they either do not believe or do not like. It is not until around the age of 6 or 7 years that children become able to separate the form of a sentence from its content, and identify sentences as acceptable or not, solely on linguistic grounds.

Given that younger children tend not to evaluate sentences on grammatical grounds in a sentence acceptability task, it is difficult to determine from studies using this method the age at which grammatical awareness develops. Although asking children to justify their responses provides information about the basis on which such judgements are made, it does not provide an adequate assessment of children's grammatical awareness, since some children may not be articulate enough to communicate the reasons for their judgements. To avoid eliciting judgements or justifications of judgements from children, Pratt et al. [in press] employed a procedure that was designed to direct children's attention toward grammatical considerations rather than content-based ones. In the study children were presented with the task of correcting ungrammatical utterances produced by puppets. To avoid eliciting judgements from children, all items were unacceptable, and children were told in advance that all items required correction.

Two groups of children, with mean ages of 5;5 and 6;4, were presented with ungrammatical sentences resulting from either word-order changes (e.g., "Teacher the read a story") or morpheme deletion (e.g., "Yesterday John bump his head"). The results indicated that when presented with ungrammatical sentences containing morpheme deletions, both groups of children performed well, with each scoring above 90% correct. However, for the word-order violations, the scores of both groups were much lower, and the performance of the older children was significantly better than that of the younger ones (48% for the 5-year-olds, as compared to 77% for the 6-year-olds).

A possible explanation of the differences in performance between the two types of sentence is that word-order changes lead to more fundamental changes in the underlying syntactic representation of sentences than do morpheme deletions. That is, in contrast with the morpheme deletions, word-order changes frequently render the original sentence meaningless and therefore require greater effort to return the sentence to an acceptable form. Moreover, it is possible that in the morpheme correction task the children may have spontaneously edited out the minor violations in the sentences presented to them in the same way that hesitations and false starts are edited out when listening to speech [see Clark and Clark, 1977]. Although this was not always the case, as some children indicated awareness of the violations by stressing the change in the corrected version, it may well have contributed to the high level of performance on this task. Consistent with this suggestion is Menyuk's [1969] finding that a certain degree of metalinguistic development is required to *inhibit* normalizations of sentences with slightly deviant syntax in a sentence repetition task. The word order correction task may therefore provide a more accurate assessment of form awareness in children, since they have no choice but to consciously manipulate the structural features of the sentence in order to return it to a grammatically acceptable form.

Related Research

In addition to research on children's acceptability judgements, there has also been some work on their judgements about whether or not pairs of sentences are synonymous, and whether certain phrases or sentences are ambiguous. Both types of judgement are rather demanding in terms of the mental operations required. Synonymy judgements, for example, require not only a comparison of the underlying semantic representations of both of the sentences in a pair, but also a comparison of their superficial forms. Both comparisons are required because synonymous sentences have essentially the same meaning but different superficial forms (e.g., "The bear chased the man" versus "It was the man that the bear chased"). If both the superficial and semantic representations of the two sentences match, the two are not synonymous but are two occurrences of the same sentence. And if neither the superficial nor the semantic representations match, the two sentences are nonsynonymous.

For ambiguous sentences, there is a match at the superficial level but a mismatch at the semantic level. An ambiguous sentence has a single superficial form but two (or more) meanings. Two types of structural ambiguity are possible -- surface-structure ambiguity, in which the words comprising the sentence may be grouped differently, giving rise to differences in meaning (e.g., "[[Tall boys] and girls] like to play" versus "[Tall [boys and girls]] like to play"); and deep-structure ambiguity, in which the two meanings associated with a sentence are not a property of surface form (e.g., "The shooting of the soldiers was awful", where the soldiers performed the act of shooting under one interpretation, and were shot under the other).

Children's judgements of synonymy have been studied by Beilin et al. [1975] and Hakes et al. [1980]. Beilin et al. examined active, passive, subject-cleft, and object-cleft sentences. Comprehension was assessed with a picture selection task, and synonymy judgements with a direct question task in which two sentences were read to a child, who was then asked whether they meant the same thing or different things. The children studied ranged from 4 to 7 years in age. The comprehension data showed that all age groups performed at better than chance on the active, passive and subject clefts, though even the oldest children did not perform well on the object clefts. For the synonymy task, the 4-, 5-, and 6-year-olds performed at or below chance on the active-passive pairs, with only the 7-year-olds performing above chance. For the subject- and object-cleft pairs, performance was at or below chance for all age groups, most likely reflecting the poor comprehension of the object clefts. These data suggest that correct synonymy performance emerges rather later than correct comprehension performance.

Using a task that was perhaps more suitable for children, Hakes et al. obtained synonymy judgements for a large number of sentence pairs with a variety of syntactic structures. Children between the ages of 4 and 8 years were introduced to a game involving two small toy animals, a dog and a turtle. Each child was told that the animals liked to play a game in which the dog would first say something and then the turtle would say the same thing but in a different way. After several demonstrations, the experimenter told the child that the turtle would occasionally try to trick the dog by saying something that was really different, something that did not mean the same as what the dog had said. Following each time the dog and turtle talked, the child was asked to say whether the turtle was answering the way he should or being "tricky".

Examination of the data revealed a different pattern of results for the synonymous and nonsynonymous sentence pairs. Even the 4-year-olds performed better than chance on the nonsynonymous pairs, but on the synonymous pairs their performance was significantly worse than chance. Performance on both groups of sentences improved with age, with rather dramatic improvement on the synonymous pairs occurring around age 6 years. Hakes et al. interpreted these results as reflecting a change in the basis on which the children were making synonymy judgements. They concluded that the younger children were judging solely on the basis of the forms of the sentences, without considering their meanings. This would explain why they were performing above chance on the nonsynonymous pairs, which differed in both form and meaning, while performing well below chance on the synonymous pairs, which differed in form only.

The development of children's ability to judge sentences as ambiguous has been studied by Kessel [1970] and Shultz and Pilon [1973]. Kessel asked children aged between 5 and 10 years to select the pictures depicting verbally presented ambiguous sentences. Two pictures out of a set of four represented each sentence's alternative meanings. Kessel found that only the older children, beginning with the 8-year-olds, could detect both meanings of structurally ambiguous sentences. Similar results have been obtained by Shultz and Pilon. Such results are not surprising, given the complexity of the task. After having assigned one structure and meaning to the sentence, the child must attempt to undo or ignore what he has done, and discover one additional structure and meaning that is distinct from the first.

Awareness of structural ambiguity has also been examined in studies that investigated the development of children's appreciation of riddles [e.g., Hirsh-Pasek et al., 1978; Shultz, 1974]. They typically show that a good understanding of structural ambiguity does not appear until around 10 to 12 years of age. However, such studies may not provide an accurate measure of linguistic ambiguity detection, since the cognitive demands of the riddle comprehension task are considerable [see Fowles and Glanz, 1977].

Summary and Conclusions

The available literature suggests that the ability to reflect upon the internal grammatical structure of sentences, as measured by tasks involving judgements of acceptability, synonymy, and ambiguity; sentence discrimination; sentence correction; and riddle comprehension, emerges later than the ability to comprehend sentences. The data from these studies reveal rather striking developmental changes between the ages of 4 and 8 years, lending some support to the notion that a developmentally distinct kind of linguistic functioning emerges during middle childhood.

This conclusion, however, must be tempered by the observation that findings vary considerably from study to study, suggesting that children's performance is highly sensitive to the nature of the tasks they are asked to complete. As mentioned earlier, many investigators use tasks which may well exceed the processing and/or memory capacities of young children, thus running the risk of obtaining data that may reflect other than metalinguistic abilities. More importantly, many of the tasks used to study the development of form awareness in children require *explicit* judgements on the part of the child. For those children who fail to provide correct judgements, it is unclear whether they are unable to reflect upon the structural features of language, or are merely unable to articulate and convey their awareness.

Recent research on spontaneous repairs in children's speech has shown that even very young children are able spontaneously to correct, or repair, their utterances. Children may therefore be aware of some aspects of language from an early age [Clark and Andersen, 1979]. However, whether such self-corrections of ongoing speech require deliberate reflection on the internal grammatical structure of sentences, or simply a check for a match between the intended meaning and the meaning of the utterance actually produced, remains unresolved [see Tunmer and Herriman, this volume]. Generally the research into children's syntactic awareness suggests that they develop the ability to consciously reflect upon the structure of sentences from age 5 onwards. But before any firm conclusions can be drawn about the age at which form awareness emerges in children, further attention will need to be paid to the use of appropriate techniques and procedures for assessing children's ability to reflect upon the syntactic and semantic properties of the sentences they produce and comprehend.

2.4 Pragmatic Awareness in Children

Chris Pratt and Andrew R. Nesdale

Pragmatic awareness is concerned with the *awareness* or *knowledge* one has about the relationships that obtain *within* the linguistic system itself (e.g., across different sentences) and with the relationships that obtain *between* the linguistic system and the context in which the language is embedded (e.g., speaker's ability to match his utterance to suit the listener's previous knowledge and current perspective). Although treated as another component of the general term metalinguistic awareness, it is important to note from the outset that pragmatic awareness incorporates aspects that extend beyond linguistic considerations. Thus whereas areas such as phonological awareness and syntactic awareness are confined to knowledge concerning specific aspects of the language system and are purely linguistic in nature, pragmatic awareness involves knowledge that takes into account aspects which extend beyond the components of the language system itself.

The study of *pragmatics* is concerned with the meaning of language as it is used to perform the prime function of communicating information and intentions between those participating in any given communication sequence. There are many factors that need to be taken into account when evaluating the meaning that is conveyed in communication [see Fraser, 1978; Robinson, 1972]. Those which are most frequently mentioned include *verbal* (i.e., grammatical structure, semantics), *intonational* (i.e., pitch, stress, junctures), *linguistic context* (i.e., set of shared propostions), *paralinguistic* (e.g., laughing, crying, pausing, yawning), *kinesic* (e.g., facial expressions, head and body movements, gestures and gaze and eye contact), *participant relationship* (e.g., role, age, sex) and *situation* (e.g., home, school). Although research into the communicative aspects of some of these factors has been more extensive than the others, the important point is that each has been shown to be communicative in an exchange between two or more people [see reviews by Argyle, 1969; Fraser, 1978; Robinson, 1972]. On this point, it should be emphasized that while these factors generally warrant the label *communication*, there are a variety of other non-linguistic elements which may be present and are not communicative [Wiener et al., 1972; Lyons, 1972]. In other words, whereas any non-linguistic element (e.g., even an egg-stained tie or dirty collar) can be used as a sign or cue upon which to base an inference about the communicator, non-linguistic elements are communicative only when they are socially shared signals of specific meanings, and are intentionally transmitted [Fraser, 1978].

Given the range of material that is included in the study of pragmatics, it is evident that the study of the development of *pragmatic awareness* in children will also encompass a wide range of factors. For such research should be concerned with the development of children's knowledge about the communicative process and the linguistic and nonlinguistic factors which play a role in this process. However, as will be seen from the following review of work to date, researchers who have studied pragmatic awareness in children have been selective in the issues they have considered. They have focussed *mainly* on linguistic factors and have concentrated on a small number of skills within the potential range that should be included in this area of study.

Indeed research into the development of pragmatic awareness in children has been chiefly concerned with three areas of study. The first is concerned with children's awareness of message adequacy when they are given messages that are not adequate because they are ambiguous. The second area involves research into children's ability to monitor information for its comprehensibility and examines their ability to detect inconsistencies in information presented to them. Although these two areas are both concerned with children's skills as listeners, they may be distinguished with respect to the information that is presented to the children. The third area of research is concerned with children's awareness of the need for speakers to modify their speech to suit the particular demands of the situation. There is now an abundance of work on the first of these areas, but there has been less on the second and what can only be described as a paucity of research on the third as will be seen in the sections which follow.

Awareness of Message Adequacy

In order for a given message to be adequate, it must contain sufficient information within the context in which it is uttered to convey the speaker's meaning to his listener with a reasonable degree of certainty. Research with children concerned with message adequacy has provided a great deal of information on both children's performance as communicators in producing and responding to adequate and inadequate messages, and on children's awareness of the importance of message adequacy and the problems and causes of inadequate messages. The emphasis here will be on children's *awareness* of message inadequacy, although reference will be made to their production of messages when this provides useful insights concerning children's knowledge of the communication situation.

In one study conducted by Markman [1977], children were explicitly requested to evaluate the adequacy of two sets of instructions -- one explaining a card game, the other a magic trick. Children from grades 1 to 3 participated in the study and before being told the instructions they were asked to help in determining the adequacy of instructions and in suggesting improvements in the instructions. Both sets of instructions were constructed so that one essential piece of information was missing. For example, in the card game set, reference was made to "the special card" without any instruction explaining what constituted this card. The results of the study revealed a clear age effect which showed that the first graders required much more probing about the adequacy of the instructions than the third graders. Indeed, first graders frequently had to attempt to act out the instructions before they realized that the instructions were inadequate. This suggested that 6-year-old children were not aware of message inadequacy until attempted action highlighted the problem.

Bearison and Levey [1977] and M. Pratt and Bates [1982] have also studied children's ability to appraise message adequacy. In the Bearison and Levey study, kindergarten, second and fourth grade children were asked to judge whether questions about messages were "good" or "bad". For example, the children were given the message "Jane got a bicycle for Christmas and Mary got a new coat" and asked to judge whether the question "What did she get for Christmas, a bicycle or a new coat?" was "good" (i.e., unambiguous) or "bad" (i.e., ambiguous). Subjects received 12 of these items -- six ambiguous and six unambiguous. The results revealed that the fourth grade children were able to carry out the task, detecting a mean of 5.1 of the 6 ambiguous questions. In contrast, second grade children scored a mean of 3.5 correct and kindergarten children a mean of 1.7 correct.

Pratt and Bates [1982] used essentially the same procedure as Bearison and Levey except in half of the items children were provided with pictorial support. The authors were concerned that the poor performance of the young children in the Bearison and Levey task may have been due to heavy processing demands. That is, subjects in that study were required to remember the message while they evaluated the question with respect to it. To investigate this possibility, Pratt and Bates provided 5-year-old children with pictorial representations of the messages for half of the items. They predicted that the pictures would help the children assess the questions and consequently that they would do better on these items. Although the results indicated that the children did detect more ambiguities in the pictorial condition, the mean performances for the pictorial and non-pictorial conditions were 2.21 and 1.71 (of a maximum score of 4), respectively. Thus, the children in this study, were still only operating at chance level, even in the pictorial condition.

In another study, Asher [1976] investigated children's abilities to appraise the adequacy of "clues" provided to distinguish between one of two words. For example, they were asked to judge whether *colour* was a good clue which would distinguish *yellow* from *blue*. Clearly in this example *colour* is not a good clue as it refers equally well to both items. Children from second, fourth and sixth grade were tested on 15 items. Although there appeared to be a difference with age, with the younger children scoring less, the oldest group in recording a mean of 9 correct were still only performing at chance level.

The studies by Markman, Bearison and Levey, Pratt and Bates, and Asher, discussed above, have all provided information concerning children's ability to make direct judgements of the adequacy of verbal messages. Further evidence of children's awareness of message adequacy is provided by studies which make use of the referential communication task devised by Glucksberg and Krauss [see Glucksberg and Krauss, 1967; Glucksberg et al., 1966]. There are now many studies which have used this task [see Dickson, 1981]. The ones which are discussed here provide a general account of the procedures that have been used and the findings from this research.

Typically, the task involves two participants who send messages to each other. The participants, either two children or an adult and a child, are seated at opposite ends of a table. Each participant is given an identical set of four to eight objects, or cards with pictures on them, and an opaque screen is placed across the middle of the table so that each participant can only see his own set of objects or cards laid out on his half of the table. One of the participants, the speaker, then selects a card and describes it to the other participant, the listener, so that the listener is able to select the matching card from his own set.

Cosgrove and Patterson used this procedure to study young children's listening skills, [Cosgrove and Patterson, 1977; Cosgrove and Patterson, 1978; Patterson et al., 1978]. The child's task was to select the correct card from a set of four on the basis of the adult investigator's description. The messages produced by the adult varied such that there were three different levels of adequacy. Adequate messages provided a unique description of one card only, partially adequate messages described two of the set of four cards and inadequate messages described all of the four cards. As the listener's task was to pick the one card that the speaker was describing, it was necessary in the partially adequate and inadequate conditions to seek further information before selecting a card. Otherwise, it was possible to select the wrong card. Using this technique, Cosgrove and Patterson studied children from four age groups -- 4, 6, 8, and 10 years of age. They found that 10-year-olds spontaneously asked for further information when it was necessary, but the remaining subjects simply selected a card without requesting further information to clarify the message received.

It is clear from this result that the 10-year-olds were aware of the importance of message adequacy for successful communication and were also aware of the means of resolving ambiguity -- by asking questions. However, the converse conclusions that the 4- to 8- year olds were not aware of these factors, cannot be drawn. There are a number of possible explanations why children of this age may be aware of the importance of message adequacy but fail to indicate this in the context of the Cosgrove and Patterson study. For example, younger children may simply misinterpret the task, thinking that they should guess if they are not sure. Indeed, Cosgrove and Patterson found that if they provided the children with a plan for more effective listening by suggesting to the children they ask questions if they were not sure which card to pick initially [Cosgrove and Patterson, 1977; Patterson et al., 1978] or alternatively, if the adult modelled question asking behaviour [Cosgrove and Patterson, 1978], then these variations led to an increase in question asking behaviour, except for 4-year-olds. Thus, their work suggests that children of 6 years and older are able to assess message adequacy and, once guided, use their assessment of message adequacy to ask appropriate clarification questions [Patterson and Kister, 1981]. Of course, this still leaves unresolved the question of whether 4-year-olds can detect message inadequacies. However, given that 4-year-olds failed to ask questions even when directed to do so, it is possible that they typically do not have any awareness of message inadequacy. Although the experimental task included ambiguous instructions, the 4-year-olds may not have been aware of them and their failure to ask questions may have resulted from a failure to detect those occasions when questions should have been asked. A subsequent study by Patterson et al. [1980] of the non-verbal behaviour of the 4-year-olds who participated in their 1977 study, helps to clarify this point. The authors found that although the 4-year-olds did not give any verbal indications that they realized the messages were inadequate, certain non-verbal behaviours were different. When presented with inadequate, rather than adequate messages, the 4-year-olds took longer to choose a card, exhibited more eye contact with the speaker and made more hand movements over the pictures. This pattern of behaviour suggests that these children at least perceived that the inadequate sentences differed in some way from those which were adequate. However, it is unclear whether the children recognized the inadequacy but were constrained for some reason from indicating it or whether the inadequacy was not sufficiently explicit to them to warrant comment.

Research examining children's awareness of message inadequacy when they are acting as listeners, has also been carried out by Flavell and his colleagues [Flavell et al., 1981; Beal and Flavell, 1982]. They used a modified version of the referential communication task where the children had to listen to and follow sets of pre-recorded instructions. Flavell et al. [1981] had a 12-year-old female confederate, Kiersten, record sets of instructions for building block constructions (e.g., put the yellow circle on top of the blue square). There were 23 sets of instructions altogether which each child had to follow. Seven of the instruction sets were clear and easy to follow whereas 13 were inadequate because they were ambiguous, or contained words not known to the children, or had some words obliterated by a sneeze deliberately recorded on the tape. The remaining three sets of instructions, although adequate, were complex and placed heavy demands on subjects' comprehension and memory abilities.

Flavell et al. studied the responses of kindergarten and grade 2 children (mean ages of 6;0 and 7;11 respectively). The children were first taught how to use the tape recorder and shown how to rewind and replay sections of the tape, should they wish to do so during the experimental task. The instructions for each task were then presented. After the child had completed each block construction, he was asked whether his construction would look exactly like Kiersten's and whether Kiersten had done a good job in giving instructions.

It is not possible to present full details of their findings here as Flavell et al. carried out an extensive analysis involving many comparisons. Indeed the analysis presented in their monograph probably involves too many detailed comparisons preventing a clear interpretation of all their data [see Whitehurst, 1981].

Generally, however, the results reveal a difference in performances between the two age groups. The older children showed more verbal and non-verbal signs that they had detected the inadequacies in the sets of instructions containing them, suggesting that they were monitoring their understanding more closely than the younger children. The verbal and non-verbal signs included facial expressions of puzzlement, hesitations while building, multiple building attempts, verbal comments, and attempts to seek the experimenter's help. However, although the older children did give such signs of puzzlement more frequently, it is not as though the younger children failed to notice any inadequacies. They just noticed them less frequently.

With regard to the children's responses to the two questions, the older children again performed at a higher level than the younger children. When the instructions were inadequate or complex, the older children were much more likely to indicate that their building might not look the same as Kiersten's. Under these conditions, the older children were also more likely to acknowledge that Kiersten had not done a good job of telling them how to make their construction exactly like her one when this was the case.

It is of interest to note that even in those cases where younger subjects showed obvious non-verbal and/or verbal signs of puzzlement during the building stage, they still tended to conclude that their building would look like Kiersten's and that Kiersten had done a good job in telling them. That is, younger children frequently failed to use an awareness of problems during the task to evaluate the outcome or the adequacy of the instructions. This finding is similar to that found by Cosgrove and Patterson in their investigation of listener skills discussed above, where children also failed to act even when they had some degree of awareness.

Other research which has made use of the referential communication task, but has looked more explicitly at young children's understanding of message failures resulting from inadequacies, has been carried out by Robinson and Robinson. In a series of studies [Robinson and Robinson, 1976a,b; 1977a,b; 1978a; 1980], they have explored the child's understanding of communication failure in the context of the referential communication game. The hallmark of the Robinsons' work is their "whose fault" series of questions which they ask when mismatches occur between the speaker's card and the listener's choice. For example, if the speaker describes his card which shows a man holding a *red* flower as "a man holding a flower" and the listener selects his card which fits the description but has a picture of a man holding a *blue* flower, then the child's attention is drawn to the mistake. The investigator then asks the child a series of questions to establish whose fault it was that the speaker and listener selected different cards. The results of their work revealed that when this type of mismatch occurs, children up to age 5 or 6 years will invariably blame the listener and may even suggest that the listener should listen harder. This attribution of blame occurs when the listener is the child himself, another child, the adult investigator or even puppets of Donald Duck or Mickey Mouse.

Beyond age 5 or 6, children begin to make the transition to becoming speaker-blamers. According to the Robinsons, they move through intermediate stages where they may still blame the listener but also show some appreciation that the speaker may not have given an adequate message or that there is a bit of the message missing. Finally, around age 7 or 8, children become "speaker blamers" and fully appreciate that the messages are inadequate because they lack certain details. Thus the children who fall into the "intermediate" or "speaker-blamer" categories show some or complete awareness of the need for message adequacy to fulfil the goal of successful communication.

The Robinsons have continued their work with children's understanding of communication failure in an attempt to see how children develop a full awareness of the problems posed by inadequate messages and how this understanding relates to their performance as communicators. With regard to the former, they have now produced considerable evidence to suggest that explicit discussion of communication failures enhances the development of the child's understanding of the basis of these failures. In one paper Robinson and Robinson [1981] reported both naturalistic and experimental data on this issue. Using data collected in a study of language development in the home [see Wells, 1981], the Robinsons examined the speech of the mothers to their children on occasions when there were communication failures. In each instance they coded the way the mother responded to the failure using 11 response categories that ranged from a straight forward "what?" through requests for repetition and requests for more information to an explicit statement of non-understanding. In addition to this, all the children at the time of their sixth birthday were assessed by the investigators using the "whose fault" technique. The findings showed quite clearly that only children who showed an appreciation of message inadequacy had mothers who had signalled explicit non-understanding on occasions when communication failures occurred. Children who were still listener-blamers had not been exposed to this means of dealing with communication failures.

The experimental data reported by Robinson and Robinson [1981] were based on a study in which children were assigned to one of three groups who received different types of responses to inadequate messages they produced in a referential communication game. For children in the first group, the experimenter simply guessed which card, whereas for children in the second group, the experimenter asked for more information. In the case of the third group, the experimenter explicitly discussed the problem -- "I don't know which one -- there are four like that". When subsequently tested using the "whose fault" technique, only children in the third group showed any advancement in their understanding of the reason for communication failures.

It is clear that the Robinsons' work has provided a considerable amount of information concerning young children's understanding of message adequacy, of the important changes that take place during the period from 5 to 8 years of age and of environmental factors which may influence the development of awareness of understanding about success and failure in communication. However, the use of the "whose fault" technique may not provide an unambiguous index of children's awareness of message adequacy. With regard to the children who are listener-blamers it would be inappropriate at this time to conclude from this response that they have no awareness that the messages refer to more than one referent. That is, although they blame the listener for not listening hard enough, this may be a response modelled from adults rather than a genuine belief that all the information is contained in the message. Young children are frequently told to listen more carefully to instructions which adults believe are adequate. That is, a child's failure to follow an adult's instruction is often taken by the adult as a failure on the child's behalf to listen to the instruction. Young children may model this explanation.

Further, the "whose fault" technique directs the child's attention to the participants rather than to the message. In relation to this point, C. Pratt [1982] conducted a study in which the children's attention was explicitly directed to the message. In the study, 4-, 5- and 6-year-old children were required to judge whether a puppet who had not learned to talk very well provided a good description which "told about" only one card from a display of four, or a poor description that "told about" more than one card. The 4-year-olds recorded a mean of 70% correct judgements on this task and the 5- and 6-year-olds recorded 76% and 90% respectively.

Recent work by Robinson and Robinson [in press (a)] reveals that young children have an awareness that inadequate messages refer to more than one possible card but fail to take this into account in assessing the confidence of their own choice. In one condition in their study there were three identical arrays of horses -- one for the child, and one each for two puppets, Mr Tickle and Mr Funny. The child's task was to select the correct horse for himself, Mr Tickle and Mr Funny, based on the adult's description. It had also been made clear to the child that in each round of the game only one horse from each array would be correct. For example, the adult might say that she was thinking of "the horse with the collar" when there were two horses in each array with collars -- one with a brown collar and one with a black collar. In such instances many children were aware that more than one horse fitted the description and selected different horses for themselves and/or Mr Funny and Mr Tickle. Thus these children were clearly aware at some level that the message was ambiguous. They did not, however, relate this to their confidence in their choice. When asked to indicate how sure they were about their own decision, children were frequently confident about it even when they had selected a different horse for one of the puppets. One problem here, of course, is that it is difficult to be sure on what grounds children are rating the confidence of their judgements. Children may have used the confidence rating to indicate that they were "sure" about their own choice in the sense that they did not want to change it, rather than as an indication that they knew it was definitely correct. Finally, with regard to those children who attribute blame to the speaker and reveal a full understanding of the cause of the failure -- insufficient information -- it must be noted that throughout the Robinsons' work these children do not use this awareness spontaneously to enhance their performance as either speakers or listeners. As speakers, they do not always give sufficient information in the referential communication task and often give ambiguous messages. As listeners, when presented with inadequate messages in this context, they frequently do not seek clarification spontaneously -- a finding that is in agreement with those of Cosgrove and Patterson [1977, 1978] and Flavell et al. [1981].

Furthermore, children in all blame categories will start asking questions when "metacognitive guidance" involving explanations as to why communication has failed and how to resolve failures is provided [Robinson and Robinson, in press (b)]. This suggests that even listener-blamers are sufficiently advanced to benefit from relatively short training sequences.

To summarize then, young children as listeners either fail to monitor messages for adequacy or alternatively, if they do monitor them, they fail to act upon detection of inadequacy unless encouraged to do so. Indeed it may be as late as 10 years of age before children will spontaneously seek clarification when aware of ambiguities. Despite this, however, children aged 5 years and upwards will begin to formulate questions when it is suggested they do so. Furthermore, explicit discussion of reasons for communication failure, and methods for dealing with it greatly enhance children's development of the awareness of message inadequacy. Generally, the data suggest that the 4- to 8-year-old period is one where significant changes in the understanding of communication failure and awareness of communication inadequacy take place. During this period children acquire greater degrees of awareness of the component skills involved in the referential communication process and begin to integrate and use this knowledge when faced with communication tasks. The awareness seems to begin at a very low level and may not be used deliberately by the child to guide his response. It is signalled with only a few non-verbal behaviours. However, the child soon becomes explicitly aware of the inadequacy of the message and the cause of the ambiguity.

Children's Evaluation of the Comprehensibility of Information

In the studies reported in the previous section, it would be fair to say that with very few exceptions, the instructions or messages presented to children were comprehensible. Investigators have ensured that the terms used in the messages (e.g., colour and shape terms) are all known to children and in many studies a pretest was given to check that children could produce and comprehend the terms used. Generally, the problem was that the messages were ambiguous in that they referred to more than one referent. However, a number of other studies have examined children's ability to monitor the comprehensibility of information. These studies have presented children either with information that was inherently incomprehensible or with information that contained inconsistencies.

Studies of the first type include those by Finn [1976] and Hughes and Grieve [1980] where children's reactions to bizarre questions were studied. Finn asked 5- to 8-year-old children such questions as, "Are there more Yukkays or more Oakkeys?" and, "Are there more Wugs or more Glugs?" The children responded to these questions as if they made sense -- for example, answering that, "There are more Wugs." The children were also willing to justify their answers. For example, when one 6-year-old was asked why there were more Wugs, he responded, "Because they're taller." Finn also asked the children about the questions themselves, and when asked, "Are these good or silly questions?", children would indicate that they were good questions. It would seem unlikely that children's judgements of the questions were based on the grammatical well-formedness of the questions as there is now much evidence from studies of grammatical awareness to show that, given the opportunity, children up to age 7 will opt to evaluate sentences on the basis of their meaning rather than on grammatical grounds [see Tunmer and Grieve, this volume; C. Pratt et al., in press]. Furthermore, Finn found that children did not invariably evaluate the questions as good, and of particular interest is his observation that children are extremely susceptible to paralinguistic cues, as the following dialogue with an 8-year-old reveals:

Adult: Have you seen many talking cars?
(Adult *laughs* while asking the question)
Child: No.
Adult: Is that a good or silly question?
Child: Silly.
Adult: Why?
Child: Cause.
Adult: Have you seen many talking buses?
(Normal voice)
Child: No.
Adult: Is that a good or silly question?
Child: Good.

In short, the experimenter found that if he were to laugh when asking a silly question, children would judge it as a silly question, but if he asked it in a normal voice, children would judge it as a good question. The results suggest that young children's understanding of communication is influenced by the presence of paralinguistic cues and that these cues are used by children to judge whether questions are good or silly.

Hughes and Grieve also found that children would respond to bizarre questions which did not make sense. They asked 5-year-old and 7-year-old children questions such as "Is red heavier than yellow?" and "One day there were two people standing at a bus stop. When the bus came along, who got on first?" Hughes and Grieve found that children in both age groups answered the questions and would also justify their answers. However, they did note that the older age groups were *at times* more hesitant and qualified their answers using phrases like, "I think", in their responses. Nevertheless, this work along with other evidence from studies by Wales [1974] and Carey [1978] whose experimental investigations involved some instructions which were incomprehensible, suggests that children do not evaluate the sense of what they hear. Indeed the findings suggest that children in the 4 to 8 year age range assume that what they hear will make sense and strive to make sense out of it. It should be emphasized, however, that this work does not determine whether young children are aware that information may not always make sense. Research conducted by the first author in which children were asked to judge whether each of a series of a questions made sense and could be answered or did not make sense and could not be answered, has shown that many 5- and 6-year-old children can evaluate questions judging them as "good" or "not good". The mean scores for the two age groups were 76% and 86% respectively. It seems therefore that the ability to monitor comprehensibility when it is deliberately encouraged in an experimental context is developing around 5 years of age.

Studies which have been concerned with children's ability to detect inconsistencies in information include those reported by Markman [1979, 1981]. She asked children in the 8 to 12 years age range to assess the comprehensibility of essays. The experimenter informed the children that she had written a series of short stories for children and wanted them to assess the suitability of the essays and suggest changes which would make the stories easier to understand. All passages involved contradictions towards the end of the story. In all cases these contradictions involved violations of reasoning based on the rule of inference, *modus ponens*. This rule may be stated in the following form:

If P is true then Q is true
P is true
Therefore Q must be true

However, the content of Markman's essays could all be translated into the following form which violates the rule:

If P is true then Q is true
P is true
Q is *not true*

For example, after a description of the types of fish that live in the ocean, the following information was given:

> Fish must have light in order to see. There is absolutely no light at the bottom of the ocean. It is pitch black down there. When it is that dark the fish cannot see anything. They cannot even see colours. Some fish that live at the bottom of the ocean can see the colour of their food; that is how they know what to eat. [p.646]

Even when the contradiction was explicitly stated, as in the story above, nearly 50% of the children failed to notice it -- with this percentage remaining fairly constant across the three age groups. In another condition, the implicit condition, which still involved a contradiction but less obviously stated, all but one 10-year-old child failed to notice the contradiction. In the light of these results, Markman considered several alternative explanations including the high memory load resulting from the length of the stories, the possibility of the children introducing information to resolve the contradiction, and the children's reluctance to tell an adult that there was something wrong with the story. Markman [1979] assessed these possibilities in subsequent experiments but there still remained a reasonably high percentage of children who failed to detect the inconsistencies. Consequently, Markman concluded that children fail to monitor their comprehension and do not relate information presented in different sentences. However, while there is no doubt that in this context, many children aged 8 to 12 years did not comment on the inconsistency, this cannot necessarily be taken as evidence that they are typically not aware of inconsistencies. The task is very complex and allows for many other possible explanations, even after Markman's modifications. For example, most of the stories dealt with inconsistencies in animal behaviour. Yet, one piece of information that perhaps children readily acquire is that animals behave in strange and mysterious ways. Also, as Markman [1981] points out, familiarity with the material being processed may affect comprehension monitoring. On these grounds, if the task was simplified or clearly based on most children's experiences, then it may be that children will detect and comment on inconsistencies.

Consistent with this view Harris et al. [1981] found that 8- and 11-year-old children were able to detect anomalous sentences when they read short stories that contained familiar information. The stories were constructed such that one line of text did not match the general story theme. For example, a story about two boys *playing* with a *toy* boat contained the sentence: *The two boys climb aboard.* Harris et al. measured children's reading time for each line of text and found that both age groups took significantly longer to process the inconsistent line of text. Following presentation of each of four stories, the children were asked to point to the line that did not match the rest of the story. Although Harris et al. reported a statistically significant age effect with the 11-year-old children showing a greater ability to detect the line that did not match the story, it should be noted that even the 8-year-olds scored a mean of 3.1 correct identifications. Since the stories contained more familiar information than that in Markman's stories, this result, together with work by Ackerman [1982], suggests that Markman's study does not provide a comprehensive account of children's abilities to detect inconsistencies.

A recent series of studies also shows that when children are given the task of evaluating simpler stories they are aware of inconsistencies between propositions. In one of these studies, Tunmer et al. [in press] presented 5- to 8-year-old children with eight three-sentence, stories for evaluation. A puppet was used to present the stories, four of which were consistent and four inconsistent, and the children were told that some of the stories would be silly and not make sense while others were good stories that did make sense. Children were randomly assigned to one of two conditions involving explicit or implicit story forms. In the explicit condition, the first sentence of each story stated a general principle compatible with the children's experience. This sentence corresponded to the first premise of an argument *if P is true then Q is true*. The second sentence constituted the second premise, *P is true* and the third sentence presented either the logically consistent conclusion *Q is true* or a logically inconsistent conclusion *Q is not true*. A typical example of an explicit inconsistent story is:

> When bikes have broken wheels, you can't ride them. One morning a car ran over Johnny's bike and broke the wheel. Johnny then picked up his bike and rode it over to a friend's house.

The same stories were used in the implicit condition except, in each case, the first sentence was replaced with a neutral sentence. In the example above, the first sentence was replaced with "Johnny got a new bike for his birthday". All stories were constructed so that inconsistencies did not occur within sentences. That is, children had to integrate the information across propositions in order to detect the inconsistencies. After presentation of each story by the puppet, the children were required to say whether the story was a good or silly one. They were also required to justify their judgements. Finally, after all the stories had been presented children were asked a series of probe questions to ensure that they were aware of the general principle involved in each story. For example, with regard to the bike story they were asked "Can you ride a bike with broken wheels?" All children performed well above 90% correct on these probe questions suggesting that they were familiar with the principles involved.

Overall, the results indicated that children performed better on the explicit condition than the implicit one and that there was a general improvement with age. By age 7, children correctly judged 90% of the implicit stories and 97% of the explicit stories. However, although there was a significant age effect, many of the youngest children still performed significantly above chance level. Thirteen of the sixteen 5-year-olds scored at this level. It might be concluded from these results that Markman's studies may have considerably underestimated the age at which children detect logical inconsistencies.

At the same time, however, it is probably fair to say that, compared with Markman's studies, the experiment by Tunmer et al. provided the optimum conditions for detecting children's awareness of logical inconsistencies. Aside from the stories being short and experience-based, the children expected some of the stories to be silly, and the possibility of their response being influenced by any inhibitions about criticizing an adult's stories was removed. And, the results of other studies have indicated that these sorts of factors do influence performance. For example, Nesdale et al. [1982] investigated the effects of *set* and *story length* on the detection of inconsistencies. Children were randomly assigned to one of two conditions, set and no-set. In the set condition children were told that some of the stories were silly and did not make sense. They were also given two practice items with feedback. In the no-set condition, children were simply asked to evaluate the stories. They were also given the same two "practice" items but without any feedback to illustrate the difference between good and silly stories. Length of story was varied within subjects by presenting children with four long stories and four short stories. The long stories were produced by attaching the short three-sentence stories on to the end of six sentences which developed the story theme but were neutral with respect to the consistency or inconsistency component. As might be anticipated, the results revealed an effect for age, with the 6-year-olds performing considerably better than the 5-year-olds, (80% versus 59%) and an effect for expectation with performance being higher in the set condition (71%) than in the no-set condition (54%). In addition, the length of the stories influenced the children's responses in interaction with age and expectancy. The younger children performed equally well on short (73%) and long stories (71%) when they expected inconsistencies but their performance deteriorated on long (42%) versus short stories (50%) when they had no such expectation. In contrast the older children's performance was little influenced by story length and they scored upwards of 72% correct, regardless of condition.

In a further study, C. Pratt et al. [1982] examined whether children's ability to detect inconsistencies is contingent upon the stories being based on principles within their experience. Three types of stories were used in the study. These were *experienced-based* stories which presented familiar information to children, *neutral* stories which presented new information to the children that neither matched nor contradicted experience; and *upside-down stories* which related to "upside-down land" -- a land that directly and systematically contradicted children's normal experience. Examples of inconsistent stories of these three types are as follows:

> *Experience-based*
> At night you cannot see the sun. In the middle of the night Jenny got out of bed and looked out the window. The sun was shining brightly.
>
> *Neutral*
> If it is the red stick, then it is the longest one. John looked at the different coloured sticks and picked out the red one. It was the shortest stick.
>
> *Upside-down*
> In the upside-down land the sun only shines during the night. One night in upside-down land Jenny woke up and looked out of her window. The sun was not shining at all.

The study involved 5- and 6-year-old children and within each of these age groups, children were assigned to one of the three story type conditions. The results revealed children within each age group performed equally well in the experienced-based condition and the neutral condition with 6-year-olds correctly judging over 90% of stories and 5-year-olds correctly judging 70% of stories in each of these conditions. In the upside-down story conditions however the performance dropped to 78% correct for 6-year-olds and 54% for 5-year-olds. Thus, the type of content of the stories does influence children's detection of inconsistencies in stories.

While the results of this series of studies indicate that children's awareness of inconsistencies is influenced by a variety of contextual factors, they nevertheless reveal that the ability of children as young as 6 years of age to detect inconsistencies is reasonably robust. By this age, their awareness of inconsistencies is not restricted to short stories, nor to those in which they expect inconsistencies, nor to stories that contain material contrary to their experience. Nevertheless, as Markman's research indicates, children can fail to notice inconsistencies. This seems to occur when several factors occur simultaneously in a situation which increases the processing load required to monitor for and detect inconsistencies (e.g., length of story) and which influence children's expectancies regarding the likelihood of inconsistencies occurring (e.g., type and source of story).

Children's Awareness as Speakers

Successful communication rests on the ability of speakers to take adequate account of the demands of their listeners and of the context. Messages must be geared to the situation, taking into account such factors as the listener's level of understanding, his current state of knowledge about the topic, and the number of potential referents which need to be disambiguated. Despite the potential richness of this area, however, the research carried out is sparse. Nevertheless, there are a few important studies which contribute to our understanding of young children's awareness as speakers.

Shatz and Gelman [1973; see also Gelman and Shatz, 1977] have examined the ability of young children to match their speech to suit the age of their listeners. They recorded the speech of 4-year-old children to 2-year-old children, to other 4-year-olds and to adults. Analysis of the recordings suggested that 4-year-olds are sensitive to differences between the *listeners*. Thus, the utterances directed towards 2-year-olds were shorter than the ones directed towards adults and same age peers. The speech to the 2-year-olds was also syntactically less complex involving fewer coordinate constructions and subordinate conjunctions. Other adjustments they found included the more frequent use of abstract verbs (e.g., *know*, *remember*, and *guess*) in speech directed towards adults than speech to 2-year-olds. Changes in the form of imperatives also occurred. Imperatives to the 2-year-olds were direct whereas those to adults were modified with introductory phrases such as *You're supposed to...* or *You have to...* . In considering the reasons for the speech adjustments of 4-year-olds to suit the demands of their listeners, Shatz and Gelman point out that 4-year-olds are aware of the restricted nature of the linguistic and cognitive abilities of the 2-year-olds. The evidence they provide in support of this comes from the spontaneous comments produced by 4-year-olds about their younger listeners. One child, for example, commented that the 2-year-old he was talking to always said "breege" instead of "bridge". Concern about their cognitive abilities included comments referring to the unlikelihood of the younger children being able to understand how a toy worked or, alternatively, that the younger child thought it worked in a different way. Thus, if children of 4 years of age are insensitive to message inadequacies, Shatz and Gelman's work suggests that they are at least aware of the differences between listeners of different ages and make adjustments in their speech to meet these. The authors do acknowledge, however, that such adjustments may not appear in all contexts as the topic being referred to may also influence the child's language production [see Cowan et al., 1967].

Menig-Peterson [1975] has investigated another aspect of children's speech adjustments -- the modifications of preschool children's speech to suit the listener's perspective. She analyzed children's descriptions of events in two different conditions. In one condition, children described an event (e.g., a staged accident involving the spilling of a cup of Kool-Aid) to an adult who had also witnessed the event. In the other condition, children described the event to an adult who had not witnessed it. The analysis examined the ways that children introduced *new elements* to their description (e.g., "*a* cup of Kool-Aid" versus "*the* cup of Kool-Aid") and whether references to objects were appropriately specified (e.g., "she was holding *a glass*" versus "she was holding *it*"). The results indicated that children did adapt their speech to suit the demands of the listener and took into account the listener's knowledge of the events being described.

Although the work of Menig-Peterson and Shatz and Gelman shows that young children can modify their speech to suit the listener, it does not directly examine children's conscious awareness of those aspects that influence the modifications. Clearly, while the differences in children's speech do reveal an implicit awareness of the demands of each situation, the study does not however, reveal the extent to which children can deliberately control their speech or make judgements about appropriate speech for particular situations [see Flavell, 1977].

Robinson and Robinson [1978b] have considered children's ability to control their message adequacy deliberately using the standard referential communication task. They asked children in the 5 to 9 years age range to give a very hard message so that the listener would not know which card to pick. This instruction resulted in 50% of the children in their study giving messages which were ambiguous and referred to more than one card. Furthermore, the ability to make messages more difficult deliberately was clearly related to the children's understanding of the reasons for communication failures. The proportion of children who were able to give difficult messages increased greatly from the listener-blamers to the speaker-blamers. This provides a clear link between children's understanding of the communication process and their ability to manipulate it deliberately.

There remains one other important study of children's ability to change their utterances when requested to do so by the adult investigator. This is the study by Bates [1976] on the acquisition of politeness forms. This research makes a very important contribution to the study of pragmatic awareness in children through its concern with children's understanding of the way social conventions may be embodied in language. That is, it is concerned with children's understanding of the social convention, politeness, as represented in their own speech and by their judgements of polite forms of request. In a cross-sectional study, Bates examined the ability of Italian children between the ages of 3 and 6 to ask nicely for some candy. The children were introduced to a handpuppet, Signora Rossi, and told that if they asked very nicely, Signora Rossi might give them a candy. Following each child's first request, the investigator said that Signora Rossi would give them a candy but only if they asked *even more nicely*. Thus Bates elicited evidence not only on polite forms used by children, but also on the children's ability to alter the degree of politeness. Following this task, Bates then presented the children with eight pairs of requests for candy and asked them to judge which of each pair was more polite.

Examining the results of her study with respect to two age groups, 3 to 4;6 years and 4;6 to 6 years, Bates found that the initial requests of the older children included more polite forms than the ones produced by the younger children. However, when asked to provide even nicer requests both groups of children were able to change the request forms. Nevertheless, an examination of these changes indicated that the older children controlled more devices for improving the politeness of requests. In addition to forms of politeness, including interrogative forms and the word *please*, that are recognized by adult standards, Bates also noted "idiosyncratic" forms that children adopt, including diminutives (e.g., "You give me a little one"). Many of these "idiosyncratic" forms suggest that children may in part equate politeness with likelihood of success of request and, for example, learn from their experience that asking for a little bit of something may lead to success.

The data from the judgement condition also showed that children develop a greater awareness of polite forms during the period from 3 to 6 years of age. However, even the youngest subjects were aware that a sentence with *please* in it was more polite than the same sentence without the word *please*. Following this, the appreciation that intonation influences politeness of request, develops around age 4. Finally, an awareness that the use of the conditional form and the use of formal codes of address increases politeness, develops between 5 and 6 years of age.

Bates' study therefore reveals that young children are aware of the means by which social conventions governing politeness are represented linguistically. But perhaps more important, is the fact that her work provides a good example of the insights that may be gained concerning children's awareness of the interplay between social factors and linguistic considerations. There is clearly a need for much further work in this area of pragmatic awareness.

Concluding Remarks

It is generally considered that by the time children enter school they have an adequate grasp of their language. Certainly they can communicate adequately in most situations as long as they are not faced with difficult grammatical constructions or words they do not understand. The work on pragmatic awareness reviewed in this chapter, however, indicates that although children have a good grasp of their native language, they still have a good deal to learn about the communicative process itself and the relationship between language and the context in which it is embedded.

In general the findings suggest that because children do not have a well developed understanding of the communicative process, they will often experience difficulties in resolving any problems that arise. Such problems may be the result of message ambiguities or inconsistencies which prevent straightforward processing of utterances and lead to communication failures. On these occasions, younger children may not have the necessary knowledge to sort out the communication failures. Nevertheless, it is noteworthy that although many 5- and 6-year-olds may hold mistaken beliefs about reasons for some communication failures and lack of skills necessary to rectify other failures, they are clearly ready to benefit from explicit discussions of such failures [Robinson and Robinson, in press]. They will also benefit from instructions concerning what they should do when such failures arise [Cosgrove and Patterson, 1978; M. Pratt and Bates, 1982].

In contrast to the understanding we now have of children's awareness of message adequacy, our knowledge concerning children's awareness of the relationship between language and context, and factors which may influence this, is more restricted. Although the study of children's ability to represent contextual factors and social conventions in language provides an exciting area of research which brings together aspects of children's social, cognitive and linguistic development, there has been very little research in this area to date. The challenge is clearly there for future research work to consider children's metacognitive abilities with respect to the way they develop an awareness of social relationships and conventions; the way they develop an awareness of purely linguistic devices concerned with such relationships and conventions; and the way they relate their awareness of both in reflecting upon and monitoring their use of language.

Part 3

Related Issues

3.1 Metalinguistic Awareness and Cognitive Development

Chris Pratt and Robert Grieve

In recent years it has been argued that there is an important relationship between the development of metalinguistic awareness and the development of cognitive abilities during childhood [e.g., Donaldson, 1978; Flavell, 1977; Hakes et al., 1980]. In particular, it is argued that the development of metalinguistic awareness in the preschool and early school years is closely related to the child's developing ability to reflect upon the products of his own thought processes and apply his cognitive skills in a wide range of contexts. As this may have important implications for our understanding of young children and their education, it is our intention in this paper to consider some recent arguments concerning the relationships between metalinguistic awareness and cognitive development. In order to do so, however, it is first necessary to consider recent changes that have taken place in the study of cognitive development in children, particularly those related to the preschool and early school years.

Cognitive Development

In the past, children were often viewed as miniature adults, it being assumed that children possessed the same abilities as adults but to a lesser extent [Aries, 1962; de Mause, 1974]. Thus children were thought to differ from adults in *quantity* only. With respect to intellectual functioning, this view predominated as recently as the earlier part of this century. It was thought that children scored less on tests of intelligence because their thought processes were quantitatively different from those of adults.

Although writers concerned with the education of children had argued that children were qualitatively different from adults [e.g., Montessori, 1917; Rousseau, 1762], the quantitative view of childhood largely prevailed until the views of Piaget became more widely known. When Piaget was employed in Paris in the early 1920s to standardize intelligence tests on a population of French school children, he noticed not only that younger children scored less than older children, but also that younger children's errors on certain items were predictable, and systematically wrong. That is, he noticed that young children's performances on the test items were not random, but that all young childen failed the same types of items, in contrast with older children who provided adequate performances on these items. Furthermore, when young children were questioned about their answers to the items that they had answered inappropriately, they provided justifications which appeared to reflect *qualitatively* different underlying thought processes. These early observations led Piaget to pursue the study of the development of thought in much more detail [e.g., Piaget, 1928, 1952, 1967], resulting in a theory of the growth of intelligence in the individual which holds that children's thought processes are *qualitatively* different from those of adults. Development is thus seen as consisting of a series of stages, and each stage is considered to be qualitatively different from the others.

In the present context however, we do not wish to consider all the stages of development in Piaget's theory of intelligence which covers the period from birth to late adolescence [for further details, see Piaget, 1970; Flavell, 1963; Brainerd, 1978]. Rather, we wish to concentrate on two of Piaget's stages, namely the *preoperational* and *concrete operational periods*, for these are the stages which encompass the age range -- from 2 to 11 years -- that is the focus of attention in the present paper.

As the name suggests, the preoperational stage, from about 2 to 7 years of age, is the period prior to the development of operations. For Piaget, operations are viewed as internalized mental representations of actions, that obey logical rules of organization. It is the development of these operations that leads to integrated and logical thought. Thus Piaget argues that the child in the preoperational stage lacks the ability to think in a logical manner. His thought processes are held to be "intuitive" and independent of each other, and as a result, the young child fails to complete certain tasks in the way that older children or adults would. Although there are without doubt positive developments during the preoperational period, including the development of language and social relationships, the period is generally viewed as one of negatives [Brainerd, 1978]. This can be illustrated by briefly considering some of the tasks which Piaget and his colleagues developed to show the distinctions between preoperational and operational thought. Here we will consider children's performance on tasks concerning conservation, class inclusion, transitive inference, and egocentrism.

In tasks concerned with conservation, the child's appreciation of the principle of invariance is considered. For example, in a conservation of number task [Piaget, 1952], the child is presented with two rows of objects set out as in Fig. 1(a). The child is first asked whether there is the same number in each row, and once the child agrees there are, the adult then alters the length of one row of objects by spreading them out as the child watches. This results in an array as in Fig. 1(b). At this stage the child is again asked to make a judgement about the relative quantities in each row: are there more Os, or more Xs, or is there the same number of each? In this standard situation the preoperational child will indicate that one row has more than the other; e.g., "There's more Xs". The concrete operational child, however, provides an answer which is in keeping with the type of answer an adult would give. That is, he asserts that there is still the same number in each row, and on questioning he can normally provide an explanation for this -- "You didn't add any more"; "You only spread those ones out"; "If you pushed these back together they would be the same as before".

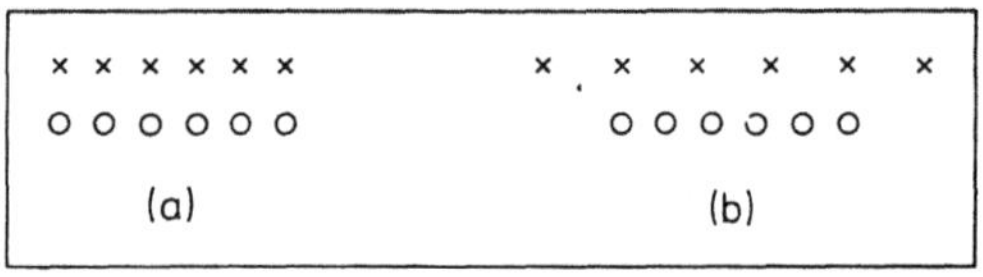

Fig. 1.a,b. Arrays presented to child in conservation task. (a) before transformation, and (b) following transformation

Thus in standard conservation tasks the equivalence of quantity is first established, then part of the array is transformed and the child asked whether the quantities are still the same or if there is more of one than the other. Initially the two quantities appear the same perceptually as well as being the same quantitatively, whereas once the transformation has been effected, the two quantities are perceptually different. Piaget argues that the young child fails the conservation task because he is perceptually misled, and cognitive operations which underlie appreciation of the principle of invariance are not yet evident [Piaget, 1952].

Similarly, Piaget argues that the preoperational child does not think logically when presented with tests of transitive reasoning [Piaget et al., 1960]. The child is first shown two sticks, A and B, and he finds that stick A is longer than stick B. Stick A is then taken away, and the child is shown stick B together with stick C which is shorter than B. Following this, the child is asked to say how A and C compare: is A longer or shorter than C? In this task, preoperational children typically say that they would need to see A and C in order to compare their lengths. Concrete operational children, however, produce the appropriate answer that A is longer than C (A>B, B>C, therefore A>C). Again Piaget's conclusion is that the young child lacks the underlying cognitive structures which enable him to make appropriate inferences and carry out the task presented.

The lack of appropriate underlying cognitive abilities is also given as the explanation of the young child's performance on tests of class inclusion [Inhelder and Piaget, 1964] and egocentrism [Piaget and Inhelder, 1956]. In the class inclusion task, the child is held to be unable to compare a class of objects with a subclass included within it. For example, if shown a class of flowers consisting of one subclass of eight primulas and another subclass of four daisies, and asked whether there are more flowers or more primulas, the preoperational child will reply that there are more primulas. He compares the two subclasses, not the subclass and the total class, as requested.

In the egocentrism task, Piaget reports that the preoperational child is not yet able to adopt a perspective other than his own. For example, shown several photographs of an object taken from various perspectives, the preoperational child can select the photograph which represents the object from his own perspective, but he cannot select the photograph which represents another perspective; e.g., a photograph taken from 90° to his right, or from 180° opposite. Often he will select the photograph which represents his own perspective instead.

Although Piaget's aim in all these tasks was to demonstrate that young children deal with problems in different ways than older children, the fact that older children complete the tasks in the same way as adults do, while young children do not, has led to the view of the young child as being *without* certain skills rather than using *different* ones. In short, Piagetian research has led to a view of the preschooler as one who cannot conserve, cannot class include, cannot make inferences, and cannot adopt perspectives other than his own.

This view of the young child largely persisted up to the mid-1970s, when a number of psychologists began to question Piaget's conclusions regarding the restricted intellectual functioning of preschoolers. In particular, publications by Gelman [1978], Donaldson [1976, 1978] and Donaldson et al. [1983] review studies which provide some evidence that young children possess a considerable range of intellectual skills, including those required to conserve, to class include, to reason, and to adopt perspectives other than their own. Often citing evidence from quite different studies, these authors advance the similar argument that young children may fail on Piagetian tasks for reasons other than the lack of the ability in question.

In investigating the conservation of length and number, for example, McGarrigle and Donaldson [1974] devised the following game. Each child was introduced to a teddy bear referred to as "naughty teddy" because he tried to spoil games. In the conservation of number task, the child was then presented with an array of the form shown in Fig. 1(a), and questioned about the number in each row to ensure that the child appreciated the equivalence. Then "naughty teddy" came and "spoiled" the game by kicking one of the rows such that the array ended up looking like Fig. 1(b). When asked whether the rows still had the same number in them, more of the younger children gave conserving responses than they did when presented with the original Piagetian task. McGarrigle and Donaldson argue that this is a result of placing the task in a context which is meaningful to the young child. They argue that the original situation is not one that is encountered normally, for there is something awry about being asked the same question twice in quick succession when nothing has changed related to the topic of the question. Yet this sort of question repetition takes place in the Piagetian conservation task. The child is first asked about the quantitative equivalence of the two rows, then he is asked the same question a few seconds later, after a transformation has been made to the array that does *not* change the quantitative equivalence. The child may therefore assume that he is being asked about something other than quantity -- for example, the lengths of the rows, because that is what has been changed. In McGarrigle and Donaldson's task, however, there is good reason to ask the same question again. The aim of the game is to make sure that the two rows have the same number, an aim with which "naughty teddy" may have interfered in trying to "mess up the game". Hence, it is acceptable to ask again about the numbers in the rows after "naughty teddy" has tried to disrupt the proceedings.

Further support for the argument about the effect of asking the same question twice comes from a study by Rose and Blank [1974]. In their study of conservation, they presented the child with the initial array (Fig. 1(a)) where the child could see that the rows were the same although he was not asked if they were. Then the experimenter changed the array (Fig. 1(b)), and asked the child for the *first* time if there were the same number in each row. This minor change in the procedure led to significantly better performance amongst the preoperational children in the study. It should be noted, however, that a recent study by Neilson et al. [1983] casts some doubt on the extent to which Rose and Blank's results may be generalized to other conservation tasks. Neilson et al. found that presenting the question only once, following the transformation, did not facilitate children's judgements in tasks of conservation of length, weight and liquid.

Since McGarrigle and Donaldson published their study on conservation accidents, there has been a series of studies which have also investigated children's conservation abilities in similar contexts [Dockrell et al., 1980; Hargreaves et al., 1982; Light et al., 1979; Miller, 1982]. While debate continues on precisely how the preoperational child's performance is to be interpreted, several studies suggest that young children can conserve in certain contexts at a considerably earlier age than the original Piagetian research suggested [Donaldson, 1982].

The view that young children do possess some appreciation of conservation is also advanced by Gelman, who draws on evidence from conservation training studies, and from her own work on children's number skills [Gelman, 1978, 1982; Gelman and Gallistel, 1978]. In training studies, children are presented with the standard conservation task which serves as the pretest. Children who pass the pretest are excluded from further study, while those who fail to provide conserving answers are then given some training. Following the period of training, the children are again given a standard conservation task, the post-test, to see if any now give conserving answers. If a significant number do, then the training is assumed to have been successful. (A no-training control group is of course required for comparison, to ensure that mere repetition of the test does not result in improved performance.)

Wallach and Sprott [1964] used this experimental paradigm when they attempted to train conservation of number ability in young children, by providing them with the opportunity to see that changing the length of a row did not alter its number, since the transformation can be reversed. The training involved a set of cardboard dolls and beds, and children were shown that the dolls could be put into the beds and taken out again. The adult demonstrated that there would always be the same number of dolls and beds provided the dolls were only put into or taken out of the beds. However, if either a doll or a bed was added or taken away, there would no longer be the same number of each. Following this training procedure, the post-test revealed that the training had been successful, as has been the case in other training studies involving different procedures [see for example Bearison, 1969; Roll, 1970; Wallach et al., 1967].

One problem with training studies, however, is that it is never clear which aspect, if any, of the training has had the effect, for the investigator can never be sure he is teaching the child the particular skill that he intends to train. While the intention in the Wallach and Sprott study was to show the child that the operation of changing the length of the rows can be reversed -- a factor taken to be an essential component in understanding conservation -- the problem is that it is not clear if this is what was actually conveyed to the children. However, Gelman and Gallistel argue that precisely what is taught during training is not the important outcome of these studies. They contend that the fact that children give conserving answers on the post-test with so little training (some studies involve as little as 15 minutes of training), implies that the children already have some underlying understanding of the concept. It would not be possible to train such a fundamental concept so quickly if the child did not already have a basic understanding of invariance. Indeed Gelman and Gallistel go so far as to suggest that perhaps what the training does is to convey to the child exactly what the adult is asking. Here again the assumption is that because of the unusual nature of the original Piagetian task, the young child does not understand what he is being asked, and training serves to convey to the child the purpose behind the investigator's questions.

With regard to other skills that Piaget's studies suggest are lacking in preschool children, there is now evidence to indicate that young children do possess and use these skills under certain conditions. With class inclusion, it has been shown that the great majority of 4-year-olds can complete class inclusion tasks successfully provided they understand which sets the adult is asking them to compare [McGarrigle et al., 1978]. In transitive reasoning, it has been claimed that provided the child remembers the information in the premises of the inference (that stick A is longer than stick B, and that stick B is longer than stick C), then he can complete transitive inferences successfully [Bryant, 1974; Bryant and Trabasso, 1971]. Alternatively, if the child is required to use transitive reasoning in measuring tasks which he can readily understand, again he can complete such tasks with high measures of success at an early age [Bryant and Kopytynska, 1976].

As regards the egocentricity of the preoperational child's thought, Hughes and Donaldson [1979] found that young children can successfully complete tasks which require taking another's perspective. In their study, preschool children were able to hide a doll from two toy policemen even though the child could still see the doll himself. The correct solution not only required the child to adopt and coordinate the perspectives of two policemen, but also required the child to place the doll in a position which conflicts with his own viewpoint. If the child had completed the task from his own perspective, then presumably he would have placed the doll where he could not see it. Other evidence which also questions Piaget's conclusions regarding egocentrism comes from the work of Shatz and Gelman. Shatz [1973, cited in Gelman, 1978] found that 4- and 5-year-old children would select age-appropriate toys for either 2-year-olds or other 4-year-olds. Thus when selecting toys for the 2-year-olds, the preschoolers selected ones which would not be suitable for themselves. Again in this context they were able to free themselves from their own viewpoint. Shatz and Gelman [1973] also investigated 4-year-olds' ability to talk to other children, aged either 2 years or 4 years, and adults. They found that when speaking to the children aged 2, 4-year-olds simplified their speech. It was syntactically less complex than when they spoke to same-aged peers or adults. The purpose of the message was also different. When addressing the 2-year-olds, the speech served to direct and maintain attention. In contrast, speech to adults sought information and clarification. Such adjustments would not be possible if the child were wholly bound by his own perspective.

As a result of the sort of work described above, the psychologist's perception of the young child is undergoing change. The young child is no longer always thought of as lacking certain skills, but as being *restricted in his use* of the skills he possesses. In certain contexts (including the original Piagetian tasks), the young child fails to use certain cognitive skills, while in other contexts he is able to use them. The focus of research has therefore become redirected. There has been a shift from studying what skills the young child lacks, towards considering what skills the young child possesses. Here, a fundamental question concerns differences in context. What is it about certain contexts that allows the young child to exhibit his skills, and how do these contexts differ from those where he fails to do so?

In considering this issue Donaldson has referred to two types of context -- *embedded* contexts where the child does use his skills, and *disembedded* contexts where he does not. According to Donaldson, embedded contexts are ones where the situation makes immediate sense to the child, whereas disembedded contexts involve situations that move beyond the bounds of immediate human sense and require more careful thought in order to complete the tasks successfully. Here then is a new problem. It seems that many skills have developed during the preschool years, but the child can only use these in certain contexts. The ability to use them in disembedded situations does not develop until after the child is at school. But what is it that develops which enables the child to deal with disembedded contexts? In considering the answer to this question it is suggested that the development of an awareness of the products of one's thought processes and, related to this, the development of metalinguistic awareness during the early school years, play an important role in the child's ability to complete disembedded tasks successfully.

Metalinguistic Awareness and Cognitive Development

The idea that children become more aware of their own thought during the early school years is by no means a new one. In the 1930s, the Russian psychologist Vygotsky wrote: "It is precisely during early school age that the higher intellectual functions, whose main features are reflective awareness and deliberate control, come to the fore in the developmental process" [translated and published as *Thought and Language*, 1962, p.90]. Vygotsky viewed this increasing control and awareness of thought as having a profound effect on children's cognitive development, and he further argued that language plays an important role in this development. The influence of Vygotsky's argument concerning the importance of reflective awareness and deliberate control is apparent in a number of recent writings in child development. However, as will become clear, authors vary in the importance they assign to the role of metalinguistic awareness in the development of cognitive control.

In extending Vygotsky's views, Donaldson [1978] assigns a key role to *metalinguistic development*, suggesting that it is largely responsible for the more general development of an awareness of thought. In outline, the argument is as follows. When the child enters the school system, he is typically expected to learn to read. This process, of learning how to extract meaning from symbols written on paper, increases the child's awareness of language. Donaldson argues that although the great majority of children have *used* language, they have never *thought* about it prior to learning to read. However, learning to read requires the child to think a great deal about language, and to become aware of its nature and uses. For example, language is composed of sounds represented by written symbols which combine to form words and sentences. Most importantly, according to Donaldson, the child also learns that written language is often used without immediate context, in a way that spoken language rarely is. At the preschool stage, the child typically relies a great deal on immediate context in understanding language. However, in learning to read, his frequent encounters with written language, where there is often no immediate supporting context, require him to pay greater heed to the language itself. In dealing with written language, the child now has to treat language more as an object of thought. Thus, according to Donaldson, learning to read leads to thinking about language, and this in turn plays a significant role in the child's general awareness of his thought processes: "... those very features of the written word which encourage awareness of language may also encourage awareness of one's own thinking and be relevant to the development of intellectual self-control ..." [Donaldson, 1978, p.95]. The development of intellectual self-control is of course seen as the key to using one's cognitive skills in a wide range of situations many of whose contexts may be "disembedded". These situations include the traditional Piagetian tasks and many others which have been devised by child psychologists and educationists to investigate children's abilities.

Eson and Walmsley [1980] have also argued that the development of metalinguistic awareness during childhood plays an essential role in the cognitive development of children. However, while Donaldson argues for a significant increase in children's awareness of language at the age when children learn to read -- that is, around 5 to 7 years of age -- Eson and Walmsley argue that an important development takes place around 10 to 12 years of age. Their view is that towards the end of the stage of concrete operational thought, children become more aware of language and that this awareness is related to the development of "semiformal" thought which exists prior to formal thinking. They support their argument with evidence from studies of children's appreciation of metaphor and jokes and their abilities to solve certain abstract cognitive tasks including, for example, the ability to predict which object will be located at the intersection of two classes of objects.

The difference in the age at which metalinguistic awareness is assumed to play a role in the development of thought seems to result from different interpretations of the term [see Tunmer and Herriman, this volume]. Although Donaldson at times uses the term to refer to the development of a general ability in children, parts of her argument suggest that she is only considering specific components of the awareness. In particular, Donaldson places special emphasis on children's awareness of the relationships that may obtain between the language and the context in which it is heard. Her argument is that young children often give too much weight to the context and too little to the linguistic signal. Until they learn to read, children are not aware that the same linguistic message may exist in different contexts, or that certain contexts will not give immediate support to the message. Further, as Donaldson argues that the original Piagetian tasks involve contexts which do not help interpret the linguistic signal, the development of an awareness of the differing roles that the context may play is essential for the successful completion of such tasks.

It would be misleading to conclude that Donaldson is only concerned with this one aspect of metalinguistic awareness, since she clearly does consider other aspects including children's understanding of the arbitrary nature of words. Nevertheless, the relationship between language and context is viewed as being important in her argument. In contrast, Eson and Walmsley, although also referring generally to metalinguistic awareness, were specifically concerned with a different type of awareness. They were interested in the development of an understanding of metaphor and ambiguity in children.

Although an awareness of these aspects of language also involves considering the relationship between the linguistic signal and the context (for example, an awareness that context can disambiguate a message), it requires in addition an understanding of the arbitrary nature of language. In particular, children must become aware that the same linguistic signal can have different meanings in different situations, or that in some cases the literal meaning is very likely not intended and that a metaphorical interpretation is required.

Consequently, until investigators are able to specify what is intended by the term metalinguistic awareness -- e.g., the type and level they are concerned with -- the debate on the age at which the development of metalinguistic awareness in the child may influence intellectual functioning is unlikely to be resolved with certainty. At present a possible explanation, in keeping with Clark's [1978] view, would be that metalinguistic awareness consists of a range of different skills, some of which are acquired at an earlier age than others. This would not preclude the possibility of significant *increases* in metalinguistic awareness being held to occur at particular points in development -- e.g., around 7 years of age -- prompted by a factor such as learning to read, for example. Nor would it preclude the possibility that awareness of a particular aspect of language -- the nature of metaphor and ambiguity, for example -- may not be well formed until the child is older.

In their writings, Donaldson, and Eson and Walmsley place a great deal of emphasis on the importance of awareness of language for cognitive development. The development of metalinguistic awareness is taken as a key factor in the development of the ability to reflect upon and control one's thoughts. Indeed, one may refer to their arguments as reflecting a strong metalinguistic awareness hypothesis, where metalinguistic awareness is considered to be a fundamental influence on thought. However, this view is not accepted by all. Flavell [1978, 1981] regards metalinguistic awareness as one of a number of parallel developments in children, all of which are considered under the general rubric of *metacognition*. According to Flavell, metacognition refers to an individual's knowledge and cognition about cognitive phenomena, which enables an individual to exercise conscious control over the products of his cognitive processes. However, although it is discussed in general terms, Flavell's use of metacognition effectively subsumes a set of specific functions, such as metamemory, meta-attention, metalinguistics, etc.. That is, the term encompasses an individual's knowledge or awareness of memory, of attention, of language etc. [see Pratt, 1981, for further discussion]. And whereas Flavell sees the development of meta-abilities as an important step in children's cognitive development, there is no suggestion that any one of these meta-abilities -- metalinguistic awareness, for example -- plays a more important role than any other in this development.

Further, according to Flavell, the development of all meta-abilities, including metalinguistic awareness, is thought to occur gradually over a period of years during childhood. For example, in a study of the development of metamemory by Flavell and colleagues, they concluded that "... the grades K and 1 subjects did seem to know some things about memory, but the grades 3 and 5 subjects seemed to know some things better and a number of other things in addition" [Kreutzer et al., 1975, p.52]. With reference to the child's ability to communicate, Flavell [1981] again expresses a similar though slightly more reserved view: "As to possible developmental trends, there *might* be an increase with age in the tendency to notice and attend to metacognitive experiences, and to evaluate their meaning, importance, trustworthiness and possible implications for cognitive action" [p.50, emphasis added].

An approach similar to that of Flavell can be found in the proposals of Brown [1978] and Brown and DeLoache [1978]. They, too, regard metalinguistic awareness as one of a group of meta-abilities which enhances cognitive functioning. Also like Flavell, Brown and DeLoache claim that in different contexts, different types of awareness are appropriate, though they also claim that there are certain metacognitive skills which are useful in "... practically any situation" [Brown and DeLoache, 1978, p.14]. In presenting their argument Brown and DeLoache acknowledge the importance of Vygotsky's view that the ability to reflect upon one's cognitive functioning (or metacognition as they refer to it) plays an important role in the child's problem-solving abilities. They claim that metacognitive skills are of great importance.

> The basic skills of metacognition, include *predicting* the consequences of an action or event, *checking* the results of one's own actions (did it work?), *monitoring* one's ongoing activity (how am I doing?), *reality testing* (does this make sense?), and a variety of other behaviours for *coordinating* and *controlling* deliberate attempts to learn and solve problems. These skills are the basic characteristics of efficient thought, and one of their most important properties is that they are transsituational. They apply to the whole range of problem-solving activities, from artificially structured experimental settings to what we psychologists defensively refer to as "real world, everyday life" situations. [pp.14-15]

From Brown and DeLoache's exposition of the importance of metacognitive skills, it is evident that the application of such general skills as predicting, and monitoring ongoing activity in specific contexts, is held to rely on specific metacognitive knowledge. Thus, the general skill of monitoring performance in a memory task will rely on awareness of memory functions, whereas the monitoring of one's performance as a speaker will involve awareness of linguistic considerations. Hence, although metalinguistic awareness is accorded an important role in Brown and DeLoache's view of cognitive development, through enabling the application of general metacognitive skills in certain contexts, it is not accorded the central or fundamental role in the development of metacognition. Rather, Brown and DeLoache see metacognitive abilities developing gradually during the preschool and early school years with children first making use of the skills in familiar contexts which are intrinsically interesting to them. In support of this argument they quote a study by Istomina [1975] on the development of voluntary memory processes in children. Istomina found that these voluntary processes, involving deliberate memory activity, were first used by children in familiar shopping game contexts rather than in unfamiliar formal memory tasks.

In fact, Brown and DeLoache argue that both children and adults will only use metacognitive skills when they have at least some experience with the problem-solving situation that they are facing. But they will not use these skills in a new and unfamiliar context. This is also held to explain why children appear to lack such skills.

> Novices often fail to perform efficiently not only because they may lack certain skills but because they are deficient in terms of self-conscious participation and intelligent self-regulation of their actions... [and that] children find themselves in this situation more often than do adults, and very young children may be neophytes in almost all problem situations. Thus, an explanation of why young children have such generalized metacognitive deficits is that most of our experimental tasks are both new and difficult for them. [p.13]

Hence Brown and DeLoache see metacognitive abilities developing gradually during childhood in familiar contexts, and like Flavell, they do not single out metalinguistic awareness as playing a greater role in the development of metcognition than other meta-abilities.

In a situation where some writers suppose that metalinguistic awareness has a uniquely important role in cognitive development, [e.g., Donaldson, 1978; Eson and Walmsley, 1980], while others suppose that it is but one of several metacognitive abilities which develop in parallel [e.g., Flavell, 1981; Brown and DeLoache, 1978], it might be thought that the results of empirical studies could be consulted to decide the matter. However, that is far from being the case at present. While some interesting results have emerged from a recent study by Hakes et al. [1980], showing that the extent of metalinguisitc awareness (as estimated by children's performance on three tests of metalinguistic abilities) is positively correlated with cognitive development (as estimated by children's ability to complete traditional Piagetian tasks of conservation), the study can be regarded as no more than it is intended to be, that is preliminary. It cannot permit a decision between the different positions just described, for although the aspects of metalinguistic awareness and cognitive development that Hakes et al. examined were found to be positively correlated, they did not consider causal relationships, and it is also not known if other, possibly parallel, metacognitive abilities would have exhibited positive correlations had they been considered.

One final major problem concerns the cause of the development of reflective awareness in children. Neither Flavell nor Brown and DeLoache provide specific details on this matter, although Flavell does suggest that metacognitive experiences will occur when an individual has cause to think about the situation -- for example, when there is a communication failure. However, Brown and DeLoache conclude their paper by acknowledging that although a lot has been said about what develops, little attention has been paid to the question of how development occurs. Indeed Donaldson's hypothesis is the only one which specifies the cause of development of awareness in arguing that it is the process of learning to read which leads to the development of metalinguistic awareness. However, although the hypothesis is attractive insofar as it does at least suggest a cause, it is not without problems.

If learning to read is seen as essential for the development of metalinguistic awareness, then presumably this would imply that preliterate children have no metalinguisitc awareness; that illiterate adults have no metalinguistic awareness; and that nonliterate societies lack metalinguistic awareness. Yet there is evidence that preliterate children are aware of some aspects of language [Clark, 1978; Pratt and Grieve, this volume; Slobin, 1978]. Further, it is now generally agreed that children require *some* metalinguistic knowledge in order to learn to read [Brown, 1980; Ehri, 1979; Mattingly, 1972]. Also, with regard to illiterate adults and nonliterate societies, there is again evidence for some metalinguistic awareness [Sapir, 1921; Heeschen, 1978]. Consequently, Donaldson's hypothesis will not hold if it is taken as proposing that learning to read is a *necessary condition* for the development of metalinguisitic awareness. Rather, learning to read may be regarded as a *typically sufficient condition* for a *significant increase* in metalinguistic awareness. In other words, most children in literate societies learn to read around age 5 to 7 years, and if they do so, this is typically sufficient to bring about a significant increase in metalinguistic awareness. Since the acquisition of literacy is not posited to be *essential* (i.e., necessary) to the advent of metalinguistic awareness, such awareness in the preliterate child, the illiterate adult, or in nonliterate societies, is thereby not precluded.

If Donaldson's proposal is regarded as claiming that learning to read is a typically sufficient condition for a significant increase in metalinguistic awareness, considerable problems still remain in evaluating this claim. For example, the acquisition of literacy in the child, around the ages of 5 to 7 years, typically coincides with the child's going to formal school. There, the child not only has to contend with learning to read written language, but he also has to contend with new uses of spoken language, and with the attendant new forms of social interaction in the formality of the school situation. The extent to which metalinguistic awareness significantly increases around age 5 to 7 years as a result of learning to read *versus* learning to contend with the changed social and (spoken) linguistic demands of formal schooling, is difficult to determine empirically. Immediately obvious empirical moves must be treated with caution. Suppose we tried to resolve the issue by examining the extent of metalinguistic awareness in children who had learned to read early, before going to formal school. Or, suppose we examined groups of children who had entered formal schooling at the same time, one a group of accomplished readers, the other a group of relatively non-accomplished readers. Such studies could well resolve nothing whatever, for a host of maturational, social, intellectual, background-familial, etc., factors could effectively mask the reading *versus* formal schooling factors under study. It is likely, for example, that children who have learnt to read early will come from different family backgrounds or will have attended alternative preschool systems to those who do not learn to read until they attend school. Here it might also be noted that while formal schooling involves the child in learning to deal with new social (spoken) linguistic forms, as well as learning to read, in the same period of around 5 to 7 years of age the child is also, typically, expected to begin to acquire systems of representation other than that involving language. That is, for the child, the advent of formal schooling coincides not only with the acquisition of literacy (involving a representation of linguistic concepts), but also with the acquisition of numeracy (involving a representation of numerical concepts). An empirical disentangling of the effects on children's awareness, of formal schooling and its attendant acquisition of different forms of linguistic and numerical representation, is certainly going to prove difficult.

Summary

In considering the relationship between metalinguistic awareness and cognitive development, it is first necessary to consider a change in emphasis that has recently occurred in the study of cognitive development in young children. Broadly, the emphasis has changed from delineating the cognitive skills that young children lack, towards determining the nature of the cognitive skills that young children possess. The limitations of the intellectual functioning of the young preschool child are now regarded by several investigators as consisting not of a lack of skills, but rather as an inability to deploy existing skills in what have been called disembedded tasks. Unlike embedded tasks, which are often completed by appeal to factors which are within the child's realm of immediate understanding, disembedded tasks appear to require a measure of reflection. The importance of the child's increasing reflection on, and control and awareness of intellectual functions, is widely recognised. The contribution made to this process by the child's increasing awareness of language, though, remains open to debate. Some writers, including Donaldson, accord metalinguistic awareness a uniquely important role in cognitive development, while others, including Flavell, consider metalinguistic awareness to be one of several "metacognitive" abilities which develop in parallel. Trying to distinguish between these views in empirical research is difficult, for many and varied factors need to be taken into account. Moreover, some of these factors, such as the child's transition to formal schooling and his acquisition of literacy, typically coincide.

3.2 Metalinguistic Awareness and Reading Acquisition

William E. Tunmer and Judith A. Bowey

The ability to read is a traditional criterion of academic achievement and is basic to success in almost every aspect of the school curriculum. It is a prerequisite skill for nearly all jobs and the primary key to lifelong learning. Despite its importance, however, it is well documented that not all children attending school attain full literacy in their native language. With regard to monolingual English-speaking children it is estimated that somewhere between 10 and 15 percent of school children having no apparent visual, hearing, or mental deficits encounter unusual difficulty of one kind or another in learning to read [Downing and Leong, 1982]. A significant number of children never learn to read efficiently or effectively. For those of us who can read, it is hard to understand why anyone should have trouble acquiring the skill, since once acquired reading seems so easy and natural. But even the average child does not learn to read easily. Citing the results of standardized reading achievement tests as evidence, Gough and Hillinger [1979] argue that "children almost never learn to read without instruction and even when given explicit, devoted, daily instruction, the average child learns to read very slowly, and with great difficulty" [p.4].

Why is learning to read so difficult? Until recently, reading was considered to be primarily a visual process, which led naturally to the conclusion that the major problem in learning to read was the failure to discriminate the visual representations of language, the letters and printed words. The solution seemed obvious. Beginning students needed to be trained in visual perceptual skills to facilitate reading acquisition. As a result, a variety of curriculum programmes for instruction in perceptual skills were developed, and a number are still in use today. However, following the research of Calfee [1975, 1977] and others [Hammil et al., 1974; Paradis, 1974; Rosen, 1966; Shankweiler and Liberman, 1972; Wingert, 1969], we now know that visual discrimination is not the central problem it was once thought to be, since there is no evidence to indicate that visual perceptual training transfers positively to growth in reading. In fact, research shows that when prereaders are administered "clean" tests, which take into account the role of memory and encoding, they demonstrate a level of visual competence altogether adequate for reading acquisition and appear to handle graphic symbols with letter-like distinctive features in much the same manner as adults [Calfee, 1975, 1977]. Training to improve visual discrimination, letter identification [Samuels, 1971, 1972; Silberberg et al., 1972] or both, does not appear to be a prerequisite to success in reading.

If lack of visual perceptual skill does not account for reading failures, then perhaps the method of instruction is responsible for the difficulty encountered by many children in learning to read. A question that has been the source of considerable controversy is whether children learn to read better by an approach that emphasizes decoding or one that emphasizes meaning [Chall, 1967]. Research has been conducted to compare a wide variety of beginning reading approaches, such as look-and-say, synthetic phonics, analytic phonics, "linguistic" methods, language experience, specialized orthographies such as the Initial Teaching Alphabet (ITA), and other variations on these themes [Bond and Dykstra, 1967; Stebbins et al., 1977, cited in Calfee and Drum, 1978]. The results of these studies indicate that only a small part of the variance in reading acquisition is attributable to differences in curriculum approaches: variability between teachers *within* one method of instruction is typically greater than variability *between* different methods. Thus, there is no compelling evidence that any one method of reading instruction is superior to the others.

However, as several investigators have noted [Calfee and Drum, 1978; Gough, 1972; Venezky, 1978], the "methods" investigated in these studies differ in ways that are so vague and poorly defined as to make them untestable. Venezky [1978] argues that most of the commonly used methods of teaching reading rely heavily on letter-sound learning, differing only in the timing of such instruction. This does not mean that comprehension is neglected. Rather, as Downing and Leong [1982] point out, "learning to read is not an 'either-or' of code-emphasis versus meaning-emphasis, but one of a continuum or different levels of processing" [p.3]. Most teachers, therefore, tend to be eclectic in their approach, using a combination of different reading methodologies [Weaver and Shonkoff, 1978]. It is probably for this reason that organizational and management variables appear to affect reading achievement more than the particular method used, leading Venezky [1978] to conclude that "there is no justification for a continued research emphasis on the so-called methods of reading instruction or on modified alphabets" [p.18].

A number of other factors have been suggested as being related to reading difficulties in children. Among these are deficient language comprehension and production skills, adverse environmental factors, emotional instability, deficient intersensory integration, minimal brain dysfunction, and genetic makeup [Gibson and Levin, 1975]. While some of these factors may indeed contribute to lower reading achievement and, in some cases, reading retardation, available evidence relating to these explanations does not account for the high percentage (10 to 20%) of "problem readers" among beginning students. More importantly, these suggested causes do not provide an explanation of where the child fails in the reading acquisition process. To design effective remedial treatment it is necessary to determine the specific language or cognitive deficiencies of the child that are preventing him from learning to read competently.

The essential question that remains is why some children benefit far more from reading instruction than do others. It has been suggested that the development of efficient processing strategies may be crucial in learning to read, and account for differences in performance [Gibson and Levin, 1975]. Research on process-oriented individual differences has resulted in the identification of several dimensions of "cognitive style", two of which appear to be related to the acquisition of reading skills; field dependence-independence and reflectivity-impulsivity [Denney, 1974; Hood and Kendall, 1975; Kagan, 1965; Kagan and Zahn, 1975; Samuels and Turnure, 1974; Stuart, 1967]. Both have been shown to be moderate predictors of reading success, especially at the lower grade levels during the initial learning of reading skills. Children who are more field independent and/or reflective tend to score higher on standardized tests of reading achievement.

Research also indicates that these two dimensions of cognitive style are developmental in nature. Children typically become more reflective and field independent with age. Moreover, studies that have examined the relationship between conceptual tempo and field dependence-independence have consistently found moderate correlations between the two [see Messer, 1976]. These findings and the finding that both variables have their greatest impact on reading achievement in the early stages of acquisition suggest that the initial learning of reading skills may be related to more general aspects of cognitive development. Reading acquisition may therefore need to be considered within a developmental context rather than as an isolated process. In support of this suggestion are several studies which report moderate correlations between Piagetian measures of cognitive development and standardized tests of reading readiness and reading achievement [Arlin, 1981; Brekke et al., 1973; Cox, 1976; Kaufman and Kaufman, 1972; Lunzer et al., 1976].

However, there has yet to be offered a clear explanation of how these developmental and cognitive style variables help the child in learning to read. Moreover, we suspect that the tests most commonly used to measure performance on these dimensions of cognitive style and cognitive development require a number of component skills, and that, of these, the one most directly involved in reading acquisition is *metacognitive control* [see Tunmer and Fletcher, 1981, for evidence in support of this claim]. The child faced with the problem of learning to read certainly needs to take an active role in the learning process, for, as Gough and Hillinger [1979] point out, we cannot give the child the orthographic cipher of English. Rather, he must discover it himself through monitoring the products of his own cognitive processes (see below).

In the sections that follow we argue that the kind of mental functioning involved in metacognitive operations underlies four types of metalinguistic ability, each of which is necessary for becoming a skilled reader. Guided by a view of reading as a set of independent processes, and reading development as the acquisition of several component skills [Calfee, 1977; Calfee and Drum, 1978], we advance a theoretical framework that describes the specific role metalinguistic abilities play in the reading acquisition process. An empirical consequence of the model is that the relative importance of the different types of metalinguistic ability varies according to the stage of development the child has reached and the particular subskills he has acquired in progressing from a beginning to a skilled reader. Although the exact nature of the relationship between metalinguistic abilities and the acquisition of reading skills is not known, there is sufficient evidence to suggest that the development of metalinguistic abilities may be central to learning to read. We conclude the paper with a brief discussion of the potential implications of our framework for future research and educational practice.

The Development of Metalinguistic Abilities

It has been generally assumed by most investigators that the beginning reader's knowledge of the spoken language is sufficiently well advanced that it does not contribute to difficulties in learning to read. Children of this age possess functional vocabularies and are able to discover the structure and meaning underlying spoken utterances. In fact, available research on language development during the first few years of life indicates that children's progress in learning to understand and produce utterances is so great that by the age of 5 years their production and comprehension abilities are approaching those of an adult [Foss and Hakes, 1978]. Although their vocabularies are much smaller, and will continue to grow for the rest of their lives, 5-year-olds produce and comprehend almost the full range of syntactic structures found in adult language [for exceptions, see C. Chomsky, 1969].

More recently, attention has begun to be focused on linguistic developments occurring after the age of 4 or 5 years, around the time when children begin to learn to read. Research on the nature of linguistic development during middle childhood (the period from 4 to 8 years) reveals that not only is there a continuation of earlier developmental processes, but that there emerges a new kind of linguistic functioning, which has been referred to as metalinguistic development. The latter refers to the ability to *reflect* upon and manipulate the structural features of spoken language, treating language itself as an object of thought, as opposed to using the language system to comprehend and produce sentences. Examples of such emerging metalinguistic abilities include detection of structural and lexical ambiguities, appreciation of linguistic jokes, segmentation of words into phonemes and sentences into words, separation of words from their referents, judgment of the semantic and grammatical wellformedness of word strings, detection of inconsistencies and communication failures, and so on [for reviews of available research, see chapters in this volume by Nesdale et al., Bowey and Tunmer, Tunmer and Grieve, and Pratt and Nesdale]. Donaldson [1978] states that prior to the development of linguistic awareness, "the child's awareness of what he talks *about*--the things out there to which the language refers--normally takes precedence over his awareness of what he talks *with*" [pp.87-88]. The child is "not able to pay scrupulous attention to the language in its own right" [p.70].

An examination of adult metalinguistic abilities reveals that the psychological processes involved are both logically and psychologically distinct from those involved in language comprehension and production. Consider the sentence, "The chicken is ready to eat", which has two possible interpretations:

1. The chicken is ready to eat (something).

or 2. The chicken is ready to be eaten.

Listeners typically understand such utterances without noticing whether or not they are ambiguous. Realizing that this sentence, or one like "Visiting relatives can be boring" is ambiguous *may* follow upon efforts to understand it, but the realization does not necessarily follow. Treating the language system as an object of thought is not an automatic consequence of using the system as a vehicle for communication.

The type of processing required in metalinguistic operations is also unlike that of normal sentence comprehension and production. Listeners are characteristically unaware that anything has intervened between their being aware of a speaker's voice and being aware that his message has been understood. Similarly, speakers are aware of what they want to say, but not of the structures by which the content of their messages is conveyed. As Cazden [1975] puts it, language is normally treated as transparent, something to be "looked through" rather than focused upon. Processing of this sort is generally referred to as "automatic".

Automatic processes are to be contrasted with "control" or "executive" processes, which entail an element of choice in whether or not the operations are performed, as well as relative slowness and deliberateness in the application of such operations. Control processing characterizes the kind of linguistic functioning associated with metalinguistic abilities, since the latter involve deliberately reflecting on the structural features of language *per se* by means of a conscious analytic ability. The language user, for example, does not normally notice such things as the individual phonemes and words comprising the utterance, the grouping relationships among the constituent words, or whether the utterance is structurally ambiguous or synonymous with another utterance, unless he deliberately thinks about it; that is, unless he invokes control processing to reflect upon the structural features of the utterance. The phrase "structural features" refers to the notion that a word is built up from phonemes, a sentence (literal meaning) is built up from words, and an intended (or integrated) meaning is a member of a set of interrelated propositions, of which the literal meaning is also a member.

The relationship between language processing and metalinguistic operations can be expressed in terms of a psychological model of sentence comprehension which is outlined in Fig. 1. The model is oversimplified, in that it specifies a set of linear, independent processors, with the output of each becoming the input to the next. Available evidence, however, indicates that there is a certain amount of interaction among the different mechanisms, in the form of feedback of information from later mechanisms to earlier ones [Foss and Hakes, 1978]. But even if it turns out that there is a *high* degree of interaction among the different mechanisms, this would not be incompatible with the existence of the various "way-stations" depicted in Fig. 1, which contain the outputs, or products, of the postulated mental mechanisms. For present purposes, the model is useful in providing the basis for classifying the various manifestations of metalinguistic awareness. Four broad categories emerge--phonological awareness, word awareness, form awareness, and pragmatic awareness. Phonological and word awareness refer to the awareness of the subunits of spoken language (the phonemes and words), form awareness to the structural representation of the literal, or linguistic, meaning associated with an utterance, and pragmatic awareness to the relationships that obtain among a set of propositions, which includes the literal and intended meanings as members.

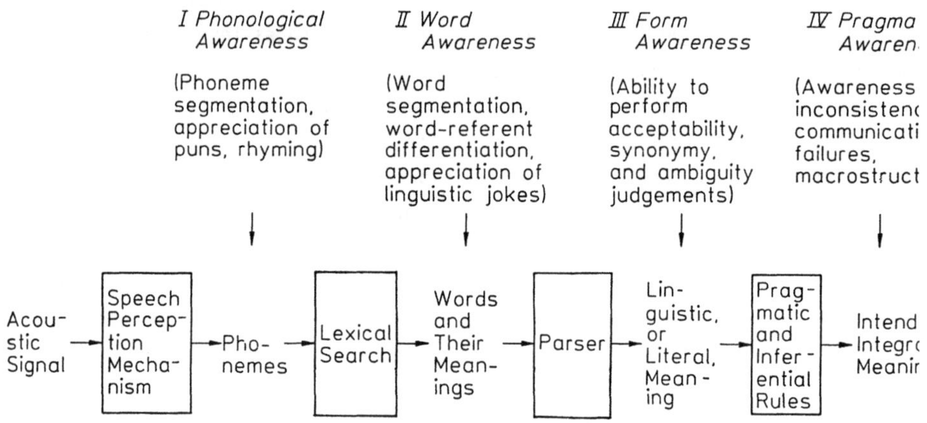

Fig.1. Types of metalinguistic awareness classified according to a model of sentence comprehension

The proposed framework suggests that the deliberate reflection and manipulation involved in metalinguistic operations are restricted to the outcome or results of prior mental processes, which themselves are automatic. Metalinguistic awareness may thus be defined in information processing terms as the ability to invoke control processing to perform mental operations on the *products* of the mental mechanisms involved in sentence comprehension, where *products* refers to the systematic phonemes, the words, the structural representations of sentences, and the sets of interrelated propositions. This definition is consistent with Piaget's [1976] views on awareness: "The results of cognitive functioning are relatively conscious, but the internal mechanisms are entirely, or almost entirely, unconscious" [p.64]. Although adult speakers are not consciously aware of, nor have *explicit* knowledge about, the subprocesses involved in sentence comprehension and production (which are represented by the rectangular boxes of Fig. 1), they have *access* to the linguistic rules embodied in such processes, where "access" is defined as the potential for forming conscious judgements based on the operation of those internal mechanisms. Such knowledge is generally referred to as *tacit* knowledge [Chomsky, 1965, 1972].

With regard to the development of metalinguistic abilities in children, recent research indicates that the emergence of these abilities may be related to a more general change in information processing capabilities that occurs during this period, the development of cognitive control, or metacognition [see Hakes et al., 1980; Tunmer and Herriman, this volume]. Metacognition (which includes metamemory, metalearning, meta-attention, metalanguage, and metasocial cognition) refers to the individual's knowledge about how he can control his thought processes to perform mental operations on the products of other mental operations; that is, it refers to "cognition about cognition" [Flavell, 1977, 1978, 1981]. Donaldson [1978] stresses the regulatory aspect of metacognition, or "disembedded thinking" to use her terminology, suggesting that children are initially not very good at exercising control over potentially useful self-monitoring cognitive strategies. She concludes that the "question of control is at the heart of the capacity for disembedded thinking which....involves sticking to the problem and refusing to be diverted by knowledge, by beliefs, or by perceptions which have nothing to do with it" [pp.93-94].

The recent work on metacognition and the observation that middle childhood is the occasion for the emergence of a wide variety of linguistic skills with the properties of metalinguistic abilities, suggest that (1) the array of diverse metalinguistic abilities that develop during middle childhood, while superficially dissimilar, are all the result of a new kind of linguistic functioning that emerges during middle childhood, the development of metalinguistic operations, and (2) these emerging metalinguistic abilities are the reflection of an underlying change in cognitive capabilities, the development of metacognition (see Fig. 2). There is some recent research pertaining to the first hypothesis [e.g., Hakes et al., 1980], but, as yet, little empirical evidence to link the development of metacognition to other aspects of linguistic development during middle childhood. However, the

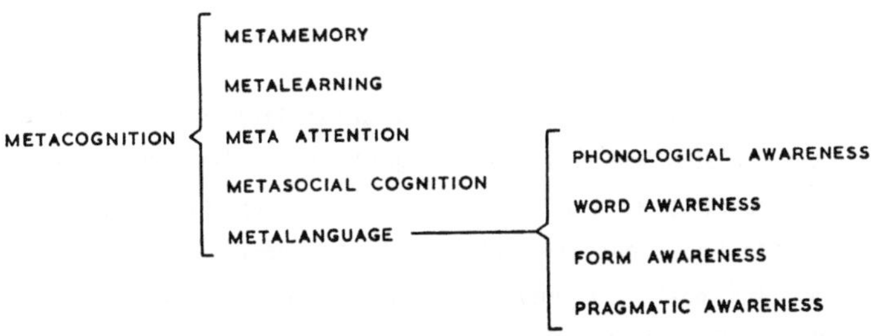

Fig.2. Hypothesized relationship between emerging metalinguistic abilities and metacognition

history of research on both metalinguistic and metacognitive abilities is relatively short. Further research is required to determine whether there exists a common denominator underlying the emergence of a broad range of cognitive and linguistic abilities during middle childhood.

Learning to Read

An important implication of the research on metalinguistic abilities concerns the problem of learning to read. Most educators and researchers view reading in relation to speech and language. As Gough [1975] argues, "knowledge of the language being read is at the heart of the reading process, and without that knowledge reading could simply not take place" [p.15]. Reading is a derived skill in that it builds upon language; the reading process is grafted on to the listening process. Accordingly, Venezky [1976] defines reading as "the translation from writing to a form of language from which the reader already is able to derive meaning" [p.6]. To learn to read, children must bring their knowledge of the spoken language to bear upon the written language. This requires the ability to deal explicitly with the structural features of the spoken language. The metalinguistic ability to reflect upon language should therefore be an important prerequisite for being able to learn to read, since without this ability the child would not be able to discover the properties of spoken language that are central to the correspondences between its written and spoken forms.

Of the four categories of metalinguistic awareness described above, we hypothesize a sequential ordering in their relative importance as the child progresses from a beginning to a skilled reader. The order is word awareness, phonological awareness, form awareness and pragmatic awareness. But prior to the application of his emerging metalinguistic abilities to the task of learning to read, the child must have some notion of what reading is all about. This we refer to as general awareness of the purpose of reading, and consideration will be given to it first.

General Awareness

Cronbach [1977] states that "a person in an unfamiliar situation must find out what to do..." [p.396], which involves "getting in mind just what is to be done" [p.398]. Bruner [1971] also describes the cognitive aspect of acquiring a new skill: "In broad outline, skilled action requires recognizing the features of a task, its goal, and means appropriate to its attainment; a means of converting this information into appropriate action, and a means of getting feedback that compares the objective sought with present state attained" [p.112]. In discussing the matter of "cryptanalytic intent", the realization of which is a precondition for learning to read, Gough and Hillinger [1979] argue that "the child must recognize that the printed message is an encoded version of a spoken one" [p.17]. Without that insight, the beginning reader can have no idea of what he is supposed to do.

Many children, however, enter reading instruction in a state of "cognitive confusion", lacking both a clear understanding of the purpose of reading and of the technical terms (*word, sentence, sound*, etc.) used by the teacher [Downing, 1979]. For example, Reid [1966], who interviewed Scottish 5-year-olds during their first year of reading, found that many of the children showed a "general lack of any specific expectancies of what reading was going to be like, of what the activity consisted in, of the purpose and use of it" [p.58]. Some "were not even clear whether one 'read' the pictures or the other 'marks' on the page" [p.60]. These findings have been confirmed and extended by Downing [1970, 1971-72].

For most children the earliest encounter with the written word arises in the situation of watching an adult read aloud. It seems likely that after repeated experiences of being read stories by an adult, the prereader comes to realize that what the adult says somehow depends upon those mysterious marks on the page in front of him. Listening to stories serves another important function in preparing the child for formal reading instruction. As Olson [1977] points out, print is not simply spoken language written down, but is a more context-independent, elaborated form of code. Unlike informal utterances, printed text "depends on no cues other than linguistic cues; it represents no intention other than those represented in the text; it is addressed to no one in particular; its author is essentially anonymous; and its meaning is precisely that represented by the sentence meaning" [p.276]. This is to be contrasted with the speech of the preschool child, which is highly concrete and bound to the specific situation. Calfee and Drum [1978] summarize the difficulties the child faces in dealing with a form of language that is independent from the situational context:

> From his earliest attempts to understand a book, the child is in the position of receiving a message where he can neither see the object of the message nor the person sending the message. Young children are not too good at the latter task [Krauss and Glucksberg, 1970]. What needs to be learned is the capability of putting one's self into the position of the message sender--or the text writer. [p.223]

In sum, children who have listened to printed text read by parents or siblings are more likely to benefit from reading instruction for two reasons; they are more familiar with the use of textlike language, and they are more apt to have a clearer concept of what reading is all about. We suspect that it is for these reasons that alphabet naming is the best single predictor of reading achievement. Although knowledge of the alphabet does not appear to be causally related to reading success [Samuels, 1971, 1972], children who can name the letters of the alphabet typically come from home environments that emphasize language activities (in addition to letter naming) that are important in preparing the child for reading instruction [Calfee, 1977; Weaver and Shonkoff, 1978]. In such environments, parents may read to their children, play word games with them, encourage them to watch *Sesame Street*, give them books of their own, take them to the library, familiarize them with the conventions of print, expose them to the technical terms used in reading instruction, and so on. Children who have not had these experiences, who come to school in a state of cognitive confusion with respect to reading, are ill-prepared to handle even the simplest of reading tasks. For these children, an initial exposure to the so-called "Language Experience" approach to teaching reading may be helpful in "bringing meaning to the text", since the materials used are based directly on the individual child's experience [see Auckerman, 1971].

Word Awareness

Earlier we mentioned that the reading process is grafted onto the listening process. As a result of an important universal characteristic of human language, its capacity to generate an infinite number of novel sentences [Hockett, 1966], it is simply impossible to have a writing system that assigns a unique symbol to every sentence of the language. It is therefore not surprising that all writing systems assign symbols to words or subunits of words (e.g., phonemes, syllables), which are finite in number. The written language must therefore map onto the spoken language at the level of the word, which means that the child's first task as he enters the reading acquisition process is to realize that one specific spoken *word* corresponds to one written word [see Biemiller, 1970, for supportive evidence].

But to do this the child must possess the logically prior notion that the flow of speech that he has been producing and interpreting unreflectingly for years is, in fact, composed of words [Donaldson, 1978]. There is, however, no simple physical basis for recognizing words in speech. A spectrographic examination of utterances spoken at a normal rate reveals that words are not separated by pauses or other obvious boundaries; there are usually no "spaces" between successive words, as in printed text. In fact, there are often larger spaces within words than between words [Foss and Hakes, 1978]. Words are highly derived, abstract entities resulting from the operation of prior mental processes [see Fig. 1].

An important question, clearly, is whether beginning readers are aware that utterances consist of words. (Note that this is different from simply using words to identify objects or produce utterances.) Available evidence suggests that they are not. Several investigators have reported that preschool children are generally unable to segment meaningful sentences or phrases into their component words [Ehri, 1975; Holden and MacGinitie, 1972; Huttenlocher, 1964; Karpova, 1966; Tunmer et al., in press], and two studies have found that awareness of aural word boundaries in beginning first graders is a significant predictor of their reading achievement [McNinch, 1974; Evans et al., 1979].

For purposes of teaching reading, it is, of course, possible to present the printed word and its corresponding spoken word in isolation, thus circumventing the segmentation problem. The difficulty here, however, is that young children experience most words in the context of other words, with their attention focused on the meanings conveyed by these spoken combinations, rather than on the words themselves [Ehri, 1979]. Consequently, many beginning readers do not fully appreciate the arbitrary nature of the relationship between the meanings of words and their phonological realizations; that is, that words are labels comprising phonological segments which attach to their referents by convention. As several studies have shown, young children experience difficulty separating words from their referents [Markman, 1976; Osherson and Markman, 1975; Piaget, 1929; Vygotsky, 1962]. Words are apparently viewed as inherent properties of objects, much like colour, shape and size.

A further difficulty is that most words depend upon the presence of other words for their meaning. This is especially true of function words such as articles, auxiliaries, prepositions and conjunctions, which are relatively meaningless when presented out of context (some are even pronounced differently when out of context; e.g., *the, a, to, and* etc.). Context dependent words such as functors may therefore be recognized not as real words by children when spoken in isolation but, rather, as meaningless nonsense syllables [see Ehri, 1975]. This suggests that with the possible exception of concrete nouns and some verbs, beginning readers may initially be limited to learning most new words when they are presented in meaningful sequences.

However, as will be discussed in greater detail in the next section, beginning readers must also discover the systematic correspondences that exist between the subunits of written and spoken words, the graphemes and phonemes, to be able to identify (i.e., decode) words that they have not seen before.

Although the grapheme-phoneme correspondences involve units other than words, there are two aspects of the mapping relationship between spoken and written language which suggest that at least some degree of awareness of words as units of spoken language is required to detect these correspondences. Firstly, research by Ehri and Roberts [1979] indicates that knowledge of grapheme-phoneme correspondences at the *lexical* level is more highly correlated with word recognition than knowledge of *single* grapheme-phoneme correspondences. This appears to stem from the fact that English orthography is primarily a system for relating phonemes to *patterns* of graphemes co-occurring within words. Secondly, research has shown that it is impossible to segment an utterance such that each acoustic segment corresponds to one phoneme; that is, there is no one-to-one correspondence between phonemes and segments of the acoustic signal (see below). Children therefore cannot be directly taught individual grapheme-phoneme correspondences, since there is no way to present a systematic phoneme like /b/ in isolation [Gough, 1972]. Rather, children must discover the correspondences by reflecting upon spoken words and their written counterparts. As Gough and Hillinger [1979] argue, "the crucial learning event occurs when the child perceives (or thinks of) a printed word at the same time he perceives (or thinks of) its spoken counterpart" [p.27].

Although the nature of the relationship between word awareness and reading acquisition is in need of further clarification, available research suggests that the metalinguistic ability to treat words as objects of thought merits consideration as a prerequisite for learning to read. To facilitate the development of the child's word concept, word segmentation games such as tapping or placing a plastic marker for each word occurring in a spoken utterance may prove to be effective, but further work is necessary to validate these suggestions. During the first stage of reading, it may also be possible to foster word awareness by using the "look-and-say" method of instruction, where words are presented in meaningful segments (but separate from their referents) and learned by sight. A continued reliance on this method, however, is likely to lead to difficulties, as we shall argue below.

Phonological Awareness

The primary assumption of the whole word, or look-and-say, approach is that each word has a specific form and configuration which makes it easy to identify and recall. Confronted with what essentially amounts to a paired-associate learning task, the child learns to read by forming arbitrary associations between selected aspects of printed words and their spoken counterparts [Gough, 1972]. Gough and Hillinger [1979] cite experimental studies and common observations as evidence that children learn to read their first words in this manner. Pre-readers, for example, learn to read lists of dissimilar words faster than lists of similar words, as it is easier to find cues which distinguish among them [Samuels and Jeffrey, 1966]. The hypothesis of selectional paired-associate learning thus predicts that the beginning reader will easily acquire a few visually distinct sight words. However, it also predicts that the child's natural strategy of associating a familiar spoken word with some feature or attribute of the word's printed form will eventually break down. Each new word will become increasingly harder to acquire because of the difficulty of finding a unique cue to distinguish it from those that have already been learned [Gough and Hillinger, 1979; Samuels, 1970]. The child will make an ever increasing number of errors, and become confused and frustrated, unless he discovers or is led to discover an alternative strategy for establishing the relationship between the written and spoken forms of the language.

What the child must come to realize is that there are systematic correspondences between the 26 letters of the English alphabet and the 44 phonemes of the English phonological system. The principle of grapheme-phoneme correspondence is a very powerful one, since a relatively small number of units can be combined to yield an extremely large set of words. To be sure, not all of the correspondence rules are simple and there appear to be a number of exceptions. However, at the level of underlying phonological representation, the irregularity is not as great as is often suggested [see Chomsky and Halle, 1968; Venezky, 1970]. Moreover, as Gough and Hillinger [1979] argue, the irregularities that occur in the system are not arbitrary; an irregular word like *of*, for example, is not pronounced like the word *elephant*. Each letter of the word *of* correctly corresponds to a single phoneme, the first a vowel and the second a fricative.

But consider the situation faced by the student who has failed to internalize the orthographic cipher of English and continues to read English words as if they were Chinese logographs or arbitrary sequences of letters. When confronted with a new word, the child is helpless, as he has no way of finding out the sound pattern (and hence, meaning) corresponding to a word he has never seen before. The child is limited to reading the particular words he has learned as arbitrary associations. By the estimates of some [e.g., Gough and Hillinger, 1979], the child will encounter more than 20,000 absolutely novel printed words by the time he reaches college, so the problems introduced by the whole word approach are potentially quite large.

A counterargument to this claim is that *context* can be used to identify new words, thus making decoding unnecessary. Research indicates, however, that context will enable the reader to predict no more than one word in four [Gough et al., 1979]. Moreover, those words that can be predicted correctly are typically high frequency function words, as opposed to the less frequently occurring but more meaningful content words [see Alford, 1980, cited in Gough and Hillinger, 1979]. That is, "the words which are predictable will tend to be those words which the child already recognizes, and the novel words which he now must recognize are exactly the ones which context will *not* enable him to predict" [Gough and Hillinger, p.13]. These considerations and the argument that the difficulty of learning a new sight word increases with each additional item, strongly suggest that the child *must* learn to decode to become a fluent reader.

To break the orthographic code, the beginning reader must figure out what phonemes go with what graphemes, which requires the ability to recognize the units of the written and spoken language. As mentioned earlier, research by Calfee [1977] indicates that pre-readers handle graphic symbols with letterlike distinctive features in much the same manner as adults. The ability to recognize graphemes and the order in which they occur does not appear to be a major stumbling block.

The same, however, cannot be said for phonemes. Recent research indicates that most 5-year-olds and many 6- and 7-year-olds are not able to perform a conscious and explicit segmentation of words into their constituent phonemic elements [Bruce, 1964; Calfee, 1977; Calfee et al., 1972; Calfee, et al., 1973; Hakes et al., 1980; Liberman et al., 1977; Rozin and Gleitman, 1977; Tunmer and Nesdale, 1982]. Several investigators have implicated this lack of awareness of the spoken language's phonemic level of analysis in the difficulties encountered by many first graders in learning to read [Calfee, 1977; Helfgott, 1976; Foss and Hakes, 1978; Fox and Routh, 1975, 1976; Gleitman and Rozin, 1977; Goldstein, 1976; Liberman, 1973; Mattingly, 1972; Savin, 1972; Wallach and Wallach, 1976; Tunmer and Fletcher, 1981]. (See also Ehri [1979] and Golinkoff [1978] for reviews of studies that have examined the relationship between phonemic segmentation skills and reading achievement.) Gough and Hillinger [1979] go so far as to suggest that lack of phonological awareness may be the most important barrier to reading acquisition yet discovered. To understand what the graphemes of the written language are supposed to correspond to, the beginning reader must be able to deal explicitly with the segmental phonological properties of the spoken language.

On first consideration, young children's inability to segment words into phonemes seems paradoxical, given that they can discriminate different speech sounds and use phonemic differences between spoken words to signal meaning differences. The difference, however, is that using a phonemic contrast to signal a meaning difference is not the same as *realizing* that the relevant difference is a phonemic difference. The perception and analysis of the speech code is done intuitively and at a subconscious level while the analysis of written language requires an explicit and conscious awareness of the relationship that exists between alphabetic shapes and phonological segments.

The difficulty for the child, however, is that there is no simple physical basis for recognizing phonemes in speech. Research on speech perception has shown that the "segmentation of the acoustic signal does not correspond directly or in any easily determined way to segmentation at the phonemic level" [Liberman et al., 1974, p.203]. It is not possible to segment a speech signal such that each segment corresponds to one and only one phoneme. Rather, any portion of the signal is likely to provide information about more than one phonemic segment. In this sense phonemic segments do not exist in the acoustic signal *per se*, but must be constructed from the signal.

For this reason we cannot directly teach the child the orthographic cipher, the set of grapheme-phoneme correspondence rules. "We cannot show him that this character goes with that systematic phoneme, for there is no way to isolate a systematic phoneme" [Gough, 1972, p.348]. In phonics, for example, each letter or letter pair (i.e., digraph) is paired not with a phoneme, but with a syllable. The child who relies solely on such rules will sound out a printed word like *bag* as *buh-ah-guh*, a nonsense trisyllable containing five phonemic segments. But if the child is to "map the printed, three-letter word *bag* onto the spoken word *bag*, which is already in his lexicon, he must know that the spoken syllable also has three segments" [Liberman et al., 1977, p.209]. Therefore, the child *must* be able to segment phonemically. Moreover, since the child cannot be directly taught the correspondence rules, he must discover them himself through his own self-monitoring metacognitive strategies. According to this view then, learning to read is a kind of problem-solving task.

Though phonics is not a method of teaching the child grapheme-phoneme correspondence rules, it does provide him with a useful means of collecting data. As mentioned earlier, Gough and Hillinger [1979] argue that the crucial learning event occurs when the child thinks of a printed word at the same time as he thinks of its spoken counterpart. To detect systematic correspondences between the written and spoken forms of the language, the child must experience or be given an abundantly large number of pairings of printed words with their spoken forms. By "sounding out" the printed word, the child is often able to hear something that closely approximates a word in his mental lexicon.

Gough [1972] suggests that the so-called linguistic method of reading instruction may be a more efficient way of facilitating the child's discovery of mapping rules. In this approach, the child is presented with a sequence of printed words in which only one letter or group of letters is varied at a time, while all else remains the same [Fries, 1963]. Since a single change in the printed text is accompanied by a corresponding change in the spoken language, the child should find it easier to discover what phonemes go with what graphemes.

For children exposed to either the phonics or linguistic approach, a lack of phonemic segmentation skills would almost certainly lead to difficulties in learning to read. It therefore seems advisable to give the child a better preparation for phoneme segmentation before reading instruction begins. This can be accomplished by exposing the child to various kinds of language activities, such as rhyming, sorting objects or spoken words by initial or final sounds, deleting a sound anywhere in the word, and so on [see Rosner, 1974; Wallach and Wallach, 1976]. In support of this suggestion is research which indicates that prior training in phonological awareness can facilitate reading acquisition [Rosner, 1971; Wallach and Wallach, 1976; Williams, 1980].

Form Awareness

Earlier we argued that the child's fundamental task in learning to read is to discover how to map the printed text onto his existing language. To accomplish this, the child must first recognize that one specific spoken *word* corresponds to one written word, which requires the metalinguistic ability to treat spoken words as objects of thought. The child then begins to form arbitrary associations between selected aspects of printed words and their spoken counterparts, a natural strategy that eventually breaks down. At this point, the child must discover that there are systematic correspondences between the subunits of written and spoken words. However, this requires the metalinguistic ability to segment spoken words into their constituent *phonemic* elements, and the metacognitive ability to detect the correspondences between these phonemic elements and the graphemic elements of written language.

Once the child has learned to decode, one might suppose that the task of learning to read is complete, since the child would appear to have succeeded in bridging the gap between the written and spoken forms of the language. However, it has long been observed that many students are able to recognize words but still cannot comprehend what they read [Weaver and Shonkoff, 1978]. Gough [1972] has suggested that slowness of decoding may account for poor reading comprehension. As words are decoded, their phonological representations are stored in short-term memory until enough information has accumulated to permit integration of the lexical entries into larger syntactic and semantic units, which are then transferred to long-term memory. But if decoding is slow, information currently in short-term memory may be lost before it is organized into larger, more meaningful "chunks" of information, and comprehension will suffer as a result.

In addition to temporal word spacing, nonautomaticity of decoding may also affect comprehension. On the assumption that comprehension during reading involves sharing a limited capacity system among various component tasks, comprehension will be adversely affected if most of the available processing capacity is engaged in nonautomatic decoding [Perfetti, 1977]. Evidence in support of this claim is provided in studies by Perfetti and Hogaboam [1975] and Hogaboam and Perfetti [1978].

For some children, however, slowness and nonautomaticity of decoding may not be the problem. It has been claimed that there are many readers who possess adequate decoding skills but read text in a word-by-word manner with list-like intonation, suggesting that decoding is a necessary, but perhaps not sufficient condition for comprehension [Cromer, 1970; Goodman, 1973; but see Calfee et al., 1976]. This has led Goodman [1967, 1968] and others [e.g., Smith, 1971] to conclude that such children lack the skills necessary to perceive text in meaningful units. According to this view, skilled reading is primarily an activity of exploiting the syntactic and semantic redundancies of language to generate hypotheses or guesses about the text yet to be encountered. "Reading is a rapid series of guesses, tentative information processing" [Goodman, 1968, p.19]. The skilled reader "perceives whole graphic phrases in an instant, processes the information and moves on" [Goodman, p.19]. Poor readers, on the other hand, are less able to make use of contextual redundancy. As Smith [1971] argues, "the more difficulty a reader has with reading, the more he relies on the visual information....the cause of the difficulty is inability to make full use of syntactic and semantic redundancy, of nonvisual sources of information" [p.221]. A consequence of this view is that reading instruction should place considerably less emphasis on decoding and more on "reading for meaning".

Although from outward appearances the skilled reader seems to go from print to meaning as if by magic, to paraphrase Gough [1972], one must resist the temptation of ascribing to such an observation the status of a "theory" of how it is done. Syntactic structure and immediate context certainly influence the *interpretation* of what is seen, as demonstrated by the disambiguation of words with multiple meanings (e.g., "I'll *bank* your cheque in the morning" versus "I'll *bank* the plane to the left", or "His *sentence* was five years and three months" versus "His *sentence* was awkwardly constructed"). Moreover, several studies have shown that preceding context can increase comprehension by facilitating the integration of new information into the reader's knowledge base [Bransford and Johnson, 1972, 1973; Dooling and Lachman, 1971]. The question, however, is whether prior context can facilitate *ongoing* word recognition [Stanovich, 1980]. That is, does prior context serve as the primary means of identifying each successive word in a sentence like the one just read? Recent research suggests not. Several studies [e.g., Forster, 1981; Gough et al., 1979; Stanovich, 1980] have reported evidence against the hypothesis-testing model of word recognition posited by such "top-down" theorists as Goodman [1967] and Smith [1971].

Rather than "sampling the text" and making sophisticated guesses as to what is on the page (which might be appropriate for skimming, but not for "reading for detail") a more efficient "strategy" for the skilled reader would be simply to read the text. When problems are encountered by the skilled reader, they typically occur at a point beyond the decoding stage, as when the reader suddenly realizes that he has not been paying attention to what was read and must reread the passage. Although predecoding readers may initially resort to the strategy of making guesses that are based on prior context, they must eventually acquire the ability to read what is actually written on the page, as research by Biemiller [1970], Juel [1980], Stanovich [1980] and Stanovich et al. [1981] suggests. To quote Gough [1972], "The good reader need not guess; the bad should not" [p. 354]. We therefore reject the view that the amount of grapho-phonemic knowledge required by the beginning reader is extremely small.

This is not to suggest that comprehension should not be stressed while decoding skills are being developed. As Downing [1979] has argued, "all instruction in subskills such as letter-sound associations should be organized so that it appears to arise quite naturally from the pupil's desire to learn the easiest and most efficient ways of getting meaning from print" [p.45]. There are numerous ways of combining decoding and comprehension instruction in the early grades. For example, students can be exposed to materials and techniques that require demonstration of comprehension (such as following directions, acting out stories, etc.) and be given the opportunity to apply their newly acquired decoding skills to the reading of enjoyable books [Weaver and Shonkoff, 1978].

Nevertheless, as mentioned earlier, some students seem to learn to decode but not to comprehend, despite the teacher's attempt to convey to the student that the purpose of reading is comprehension. Downing and Leong [1982] state that "failure in the integration process is a well-known feature of reading disability--for example, in pupils who can decode letters to sounds or printed words to spoken words but cannot integrate these skills in fluent reading" [p.35]. The present theoretical framework provides a natural interpretation of this phenomenon. To avoid the "bottleneck" problem associated with short term memory limitations, and to learn to identify words not seen before, the child must master the grapheme-phoneme correspondence rules. But to do this, the child must focus on the word, for, as argued earlier, the crucial learning event occurs when the child thinks of a printed word at the same time he thinks of its spoken counterpart. This means that in the early stages of the reading acquisition process, the child must *ignore* information that is not relevant to the task at hand. He must separate the word from its sentential context (which, as noted earlier, is not easy for young children to do) and focus his attention on discovering systematic correspondences between the units of written and spoken *words*.

But as soon as the child has fairly well mastered the grapheme-phoneme correspondence rules, he must change his strategy and put humpty-dumpty back together again. He must consciously begin to organize the text into higher-order *syntactic* groupings, since the structures of sentences are crucial to their understanding. He must therefore bring his syntactic knowledge of the spoken language to bear upon the written language, which again requires the metalinguistic ability to reflect upon the structural features of spoken language.

However, as noted earlier, it is not until middle childhood that children develop the ability to analyze and manipulate aspects of language that were previously mastered unconsciously by listening to and producing utterances in supportive contexts. With regard to the development of form awareness, Pratt et al. [in press] found that when the nature of the task was such that children were required to focus their attention on the structure of the sentence, as in a task in which they were asked to "unscramble" short sentences containing word-order violations, young children tended to perform poorly. This might explain why some children do not spontaneously use their implicit syntactic knowledge and continue to read word-by-word. A certain amount of syntactic awareness may therefore be necessary to facilitate reading comprehension, suggesting that some children may require assistance in developing this ability.

A recent study by Weaver [1979] provides some evidence in support of this suggestion. She found that instruction in the use of grammatical structure transferred positively to reading comprehension. Of particular interest is her finding that average and above average readers responded differently to training which involved a sentence anagram task and introduction to a word-grouping strategy. Weaver reports that the "above average readers spontaneously developed and used solution strategies that average readers had to be taught to use" [p.144]. Similarly, Valtin [1979] found that children with reading difficulties did not differ from normal readers in a variety of oral language measures, but in a grammar test that required the children to *consciously* apply their grammatical knowledge, the dyslexic children showed inferior results. Other studies using different methodologies have also found a positive relationship between sentence organization skills and comprehension [e.g., Denner, 1970; Resnick, 1970; Weinstein and Rabinovitch, 1971].

Further evidence in support of the hypothesized relationship between form awareness and reading comprehension comes from studies of reading achievement that have used the cloze task, where the subject is required to produce the missing words in visually presented sentences. Since the cloze task requires the subject to consciously analyze the sentence in order to select the word which is most appropriate in the surrounding semantic and syntactic context, it can be argued that the cloze task constitutes, in essence, a written form awareness task. This suggestion is supported by the finding of a positive correlation between the cloze task and an oral form awareness task involving sentence correction [Bowey, 1982]. Research indicates that performance on cloze tasks correlates positively with reading comprehension scores [Bickley et al., 1970; Ruddell, 1965], that training on cloze tasks transfers positively to reading comprehension [Kennedy and Weener, 1973], and that, of the various subtests of the Illinois Test of Psycholinguistic Abilities [Kirk et al., 1968], the Grammatical Closure subtest (an oral cloze task) correlates the most highly with early reading achievement [Newcomer and Hammill, 1975]. With regard to such findings, Juel [1980] argues that "researchers may be mistaken in interpreting ... studies which show good readers can make use of contextual cues better than poorer readers as evidence that they actually do so in normal reading" [p.375]. That is, the fact that good readers demonstrate better predictive abilities does not mean that they rely on such abilities to facilitate *ongoing* word recognition [Stanovich, 1980]. The question that remains, however, is why better predictive abilities are associated with greater reading proficiency.

One possibility is that well developed form awareness enables better readers to *monitor* more effectively their comprehension [Golinkoff, 1975-1976; Stanovich, 1980]. As Perfetti [1977] has suggested, an important factor in skilled comprehension may be what he calls "attention for language" [p.25]. In support of this view is Weber's [1970] study of first graders' oral reading errors. She found that good readers were much more likely than poor readers to correct errors that did not conform to the syntactic and semantic form of the sentence, which seems to indicate that they were more aware of when they were not comprehending the text. Bowey [1982], similarly, found that children scoring high on an oral form awareness task were better able to exploit contextual information in their ongoing oral reading performance. Such children also read more fluently, and scored higher on standardized reading tests than did children scoring poorly on tests of oral form awareness. These findings, and those discussed earlier, provide additional support to the claim that metalinguistic awareness plays a central role in learning to read.

Pragmatic Awareness

In addition to organizing words into larger structural units, children must also notice relationships that obtain *among* groups of sentences and the context in which they are embedded to fully understand what they read. They must integrate the sentences into larger structural representations, which involves, in part, combining new information with old information. Since most messages do not explicitly contain all the information necessary for effective comprehension, children must "fill in the gaps" in messages by making inferences and assumptions based on their prior knowledge. The integration of new information also depends on the degree to which the listener (or reader) and speaker (or text writer) share a *common semantic field*, or *common ground*, which is the set of propositions the listener and speaker are rationally justified in taking for granted (such as the speaker's intention, the listener's expectations, contextual and situational cues, and knowledge of intended referents).

Although the sentences of connected discourse are not structurally dependent, they are linked together by *cohesive elements* to form a unified whole [Halliday and Hasan, 1976]. Cohesive elements are semantic in nature and refer to relations of meaning within the text. A number of investigators have also attempted to identify the general organizational structures in prose. Kintsch and van Dijk [1978], for example, have developed a general theory of discourse, according to which sequences of interrelated sentences are organized into larger structural units called *macrostructures*, which represent the main aspects of a text that are stored during comprehension.

In a review of research on reading comprehension processes in good and poor comprehenders, Golinkoff [1975-1976] concludes that poor comprehenders appear to possess less ability than good comprehenders to organize text into these higher-order structural units. Recent findings from the developmental literature point to a possible explanation of the comprehension difficulties encountered by poor readers [see Dickson, 1981]. Several studies indicate that children often fail to notice when they do not understand, as in situations where there is insufficient or ambiguous information, or inconsistencies in the information provided [Cosgrove and Patterson, 1977, 1978; Markman, 1977, 1979; Tunmer et al., in press]. Other research has shown that younger children generally do not understand the causes of failure in verbal communication [e.g., Robinson and Robinson, 1977a], and cannot readily make explicit judgements of the structure of prose passages [Brown and Smiley, 1977].

Flavell [1977] suggests that these communicative behaviours may require a fairly late-developing "metacommunicative" ability, which he defines as "the ability to take a verbal-communicative message as a cognitive object and analyze it" [p.178]. A related interpretation of these findings is that younger children fail to engage in the inferential and constructive processing necessary to obtain information *about* their own comprehension. Markman [1981] suggests that "comprehension monitoring" requires the metacognitive ability to reflect on and direct one's mode of cognitive functioning, and that "without such knowledge about comprehension, comprehension itself will suffer" [p.81]. Given the clear educational implications of these studies, more empirical work is needed to determine what techniques can be used to promote comprehension monitoring and text organization in beginning readers.

Summary and Conclusions

Throughout this paper we have argued that the development of metalinguistic abilities in children is central to learning to read. A theoretical framework was advanced which describes the specific role these abilities play in the reading acquisition process. While the research evidence reviewed indicates that the four general types of metalinguistic ability (word awareness, phonological awareness, form awareness, and pragmatic awareness) are significantly related to reading achievement, further research is required to test the stronger hypothesis that the relative importance of these four types varies according to the stage of reading achievement reached by the child.

In opposition to the view that metalinguistic awareness is a *prerequisite* for learning to read, is the "interactionist" position of Ehri [1979]. Ehri and Wilce [1980], for example, claim that phonological awareness is a "consequence of as much as a prerequisite to learning to read words" [p.371]. However, their claim is not logically distinguishable from the view that metalinguistic awareness is a prerequisite only, since the statement that metalinguistic awareness is a *necessary* (i.e., prerequisite) condition for acquiring reading skills *logically* implies that learning to read is a *sufficient* condition for bringing about metalinguistic awareness. The view that metalinguistic awareness is a prerequisite skill is not inconsistent with the possibility that reading *instruction* increases metalinguistic awareness, which would explain the sharp increases in metalinguistic abilities often observed among beginning readers. As Byrne [1981] has argued, "most reading programs directly, if not deliberately, teach children about sentences and word structure" [p.86]. This observation concurs with the model of reading acquisition outlined in the present paper, where different teaching methods were explicitly identified with different aspects of metalinguistic awareness.

The stages of reading acquisition hypothesized in the model are summarized in Fig. 3. As can be seen in the figure, the transition from one stage to the next does not occur suddenly. Rather, there is considerable overlap among the various stages. But as the child enters a new stage, he must use his emerging metacognitive abilities to generate the strategies necessary to perform mental operations on the relevant structural features of spoken language. When the particular subskill becomes "automatic", to use a distinction proposed by LaBerge and Samuels [1974], the child is in a positon to move on to the next stage.

If the present model is correct, it would suggest that both decoding *and* meaning should be emphasized in instruction, but in differing amounts at different points in the reading acquisition process. Moreover, by its very nature as a stage model, it suggests a diagnostic-prescriptive approach to remediation. Children may encounter problems in reading at various stages. But, since each stage hypothesized in this model is underlain by a different type of metalinguistic ability, remediation techniques should vary depending

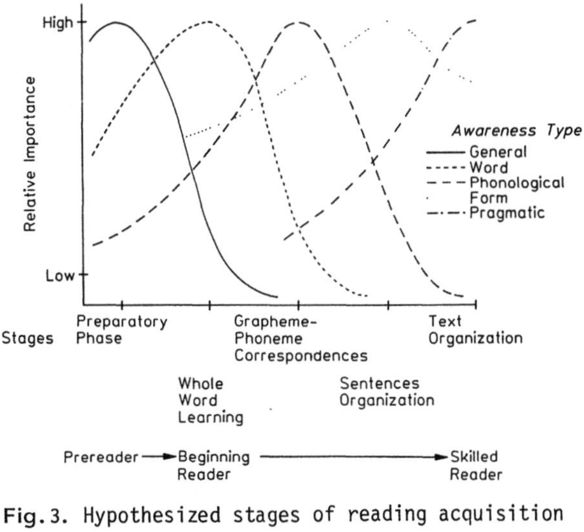

Fig. 3. Hypothesized stages of reading acquisition

on the stage of reading development reached by the child. In order to pursue these suggestions further, future research should aim to determine more precisely the nature of the relationship between the child's emerging metalinguistic abilities and learning to read, the degree to which the development of these abilities can be facilitated by training, and whether such training transfers positively to reading.

3.3 Metalinguistic Awareness and Bilingualism

William E. Tunmer and Marion E. Myhill

Research on the effects of bilingualism on children's cognitive growth and academic achievement has generated conflicting results. Older studies tend to point to negative effects for bilingualism, while more recent work indicates that, on the contrary, bilingualism may produce positive effects [see Cummins, 1979, for a review of relevant research]. One such positive effect is an increase in metalinguistic awareness. In the sections that follow, we argue that this increase in metalinguistic awareness may provide an explanation for the positive effects that fully fluent bilingualism has on academic and cognitive functioning.

The paper is divided into four sections. The first defines bilingualism and introduces the important concept of relative linguistic proficiency; the second summarizes the available research on the relationship between bilingualism and cognitive growth; the third examines the possibility that metalinguistic awareness acts as an intervening variable mediating the positive effects of bilingualism on academic achievement; and the fourth briefly discusses implications of recent findings for bilingual education.

Definition of Bilingualism

Bilingualism has been variously described as "having a non-native relative living in the home" [Lee, 1932]; "native-like control of two languages" [Bloomfield, 1946]; "alternately using two languages" [Weinrich, 1953]; and "being able to complete meaningful utterances in two languages" [Haugen, 1956]. This lack of consistency in the definition of bilingualism is probably one of the primary sources of the conflicting findings among studies concerned with the effects of bilingualism on cognitive growth. This is especially true for studies which have claimed to demonstrate the negative effects of bilingualism. As De Avila and Duncan [1979] argue, "the failure to operationalize and control for linguistic proficiency has resulted in the widespread belief that bilingual children must inevitably face a linguistic handicap with the ultimate effect of lowering both intellectual and academic performance" [p.4].

The major problem with most early attempts to define bilingualism was the failure to allow for varying *degrees* of bilingualism. As Hornby [1977] points out, "bilingualism is not an all-or-none property, but is an individual characteristic that may exist to degrees varying from minimal competency to complete mastery of more than one language" [p. 3].

While some bilinguals are fully fluent, or "balanced", in both languages, most tend to be more fluent in one language than the other; that is, one language is dominant. The question that must be answered, however, before any conclusions can be drawn about the possible effects of bilingualism or bilingual education programs on cognitive growth, is to what extent the speaker is fluent in the *non-dominant* language (or in both languages, if the speaker is not fully fluent in either). Recent attempts have been made to systematize these variations by rating oral language proficiency in both languages on a scale, such as from 1 (low) to 5 (high).

An example of this approach to measuring bilingualism is provided by De Avila and Duncan [1977], who developed the Language Assessment Scales (LAS). The LAS comprises two individually administered assessment measures of oral language proficiency in English and Spanish. Each test provides an overall level of proficiency based on the student's performance on various subscales. (For a detailed discussion of the difficulties involved in developing reliable measures of oral language proficiency in young children, see De Avila and Duncan [1978].) The five levels are defined as follows (the definitions are the same for both the Spanish and English tests):

1. Non-speakers with total linguistic deficiencies; ranges from speakers who are hearing the language for the first time to those who can produce a few one-word utterances.
2. Non-speakers with apparent linguistic deficiencies; speech is limited to two-word phrases and occasional simple sentences.
3. Speakers with limited speaking ability; speech contains complete sentences, but with systematic errors in syntax and repeated use of code-switching (borrowing of words from another language).
4. Near-fluent speakers; speech contains only occasional errors.
5. Totally-fluent speakers; speech sounds like that of a native speaker of the language.

Each language version of LAS provides a separate, composite score (level) for each language. This makes it possible to determine the *relative linguistic proficiency* (RLP) of English to Spanish for each child who is given the LAS [Duncan and De Avila, 1979]. Since there are five levels for each language version, there are a total of 25 possible RLP combinations. These combinations provide the basis for developing operational definitions for the different categories of bilingualism.

For example, *fully fluent, or balanced, bilinguals* could be defined as those children achieving RLP scores of 4/4, 4/5, 5/4, or 5/5 (the digit to the left of the slash represents the proficiency in English, the digit to the right, proficiency in Spanish). *Partial bilinguals,* who are proficient in one language, but limited in the other, could similarly be defined as those achieving RLP scores of 5/3, 4/3, 3/5, or 3/4. *Monolingual* speakers would include children possessing native-like fluency in one language but little or no ability in the second; that is, those with RLP scores of 5/1, 5/2, 4/1, 4/2, 1/5, 2/5, 1/4, or 2/4. Two kinds of speakers with deficiencies in both languages could be identified, those categorized as *limited bilinguals* (3/3, 3/2, 2/3) and those categorized as *limited monolinguals* (3/1, 1/3). *Late language learners* (2/2, 2/1, 1/2, 1/1) would be those children who demonstrate virtually no proficiency in either language, a condition which may be related to gross language disorders or mental retardation.

A definition of bilingualism which involves a categorization scheme such as this, while straight-forward, is clearly more complicated than the definitions of bilingualism cited earlier. However, as will become clear in the remaining sections of the paper, such additional complexity is essential if we are to advance our understanding of the relationships between (the different types of) bilingualism and cognitive growth.

Effects of Bilingualism on Cognitive Growth

As noted in the introduction, the early research on bilingualism indicated that bilingualism was associated with a general intellectual deficit relative to monolingualism. The bilingual child seemed not to have the conceptual or linguistic skills of the monolingual child. These findings led to the generally accepted view, which prevailed for many decades, that bilingual children suffered from "mental confusion" and "language retardation".

In a comprehensive review of this research, however, Peal and Lambert [1962] found that most of these studies contained serious methodological flaws. For example, several studies attempted to determine whether monolingual and bilingual children differed in intelligence by administering standard intelligence tests to both groups, but without controlling for socioeconomic status. Since socio-economic status is known to affect measured intelligence, the lower intelligence scores obtained for the bilingual children may simply mean that these children tended to come from lower socioeconomic groups. Another possible explanation for the differences is that the intelligence tests used in many of these studies were constructed and standardized in a culture different from the one to which the bilingual children belonged. That is, the reported differences may simply reflect the cultural bias of the test instruments used. Finally, and perhaps most importantly, few studies included an adequate definition of bilingualism. Many of the children who were classified as "bilingual" may, in fact, have been either *partial* bilinguals, with RLP's of 5/3, 4/3, 3/5, or 3/4, or *limited* bilinguals, with RLP's of 3/3, 3/2, or 2/3. This is an important consideration, since recent linguistic theories (to be discussed below) have postulated that some forms of bilingualism produce *positive* cognitive effects, while others produce *negative* effects. Thus, in order for studies on the effects of bilingualism to yield interpretable results, it is essential to know the *type* of bilingualism involved.

Peal and Lambert [1962] provided the first major challenge to the view that bilingualism was associated with inferior intellectual functioning. In a large-scale study in Canada, they studied the performance of French/English bilingual and English monolingual children on a battery of tests. Unlike most previous studies, the two groups were carefully matched for social class, educational opportunities, age, and degree of bilingualism. Contrary to the bulk of earlier research, they found that the bilingual children scored better than the monolinguals on both verbal and nonverbal tests of intelligence. On the basis of their findings, Peale and Lambert concluded that the bilingual child appears to have "a mental flexibility, a superiority in concept formation, and a more diversified set of mental abilities" [p.20]. Subsequent studies in different settings have confirmed these general conclusions [Bain and Yu, 1978; Cummins and Gulutsan, 1974; Liedke and Nelson, 1968; Torrance et al., 1970].

Consistent with these findings are the results of several studies conducted within the context of total immersion programs [Cohen, 1974; Lambert and Tucker, 1972; Swain, 1978]. In immersion programs, which normally begin at the kindergarten level, children are instructed entirely in a language which is different from their home language, and which they do not speak on entering the program.

One of the first carefully planned and evaluated immersion projects was conducted in St. Lambert, a community near Montreal, Canada [see Lambert and

Tucker, 1972]. Monolingual English speaking children were instructed entirely in French in kindergarten and first grade, with English instruction gradually introduced in the later grades. At the end of each school year the children were administered tests which measured various French and English language skills, mathematical concepts, and cognitive flexibility. The tests were also given to French and English monolingual control groups attending standard monolingual schools. Initially the immersion children lagged behind the control groups in the development of their language skills, although they did just as well on arithmetic tests in both languages. However, by the time they had reached the higher elementary school grades, the children had achieved a very high degree of functional bilingualism with no detrimental effects on their first language or mathematical skills. Moreover, the immersion children scored higher than the control groups on tests of cognitive flexibility. Other studies [e.g., Swain, 1978] have reported that English speaking immersion students performed better than controls on tests of English skills, even though they had received considerably less instruction in English.

While these studies suggest that a switch from one language at home to another at school can result in high levels of functional bilingualism and academic achievement, the opposite has been found with Spanish speaking Chicano children and other language minority children in the United States [U.S. Commission on Civil Rights, 1971]. The poor academic performance of these children cannot be attributed solely, or even primarily, to the effects of lower socioeconomic status and racial discrimination, since Black children do not perform nearly as poorly. Attempts to explain the lower educational attainment of Chicano children immersed in regular all English programs have focused on cognitive style differences, socio-cultural and attitudinal factors, and linguistic factors.

Cognitive style generally refers to individual variation in preferred modes of perceiving, remembering, and thinking, and includes such dimensions as field dependence - independence and reflectivity-impulsivity (or conceptual tempo). Research has been reported which suggests that certain ethnolinguistic minority groups have preferred styles of learning which are different from the dominant culture [e.g., Ramirez et al., 1974; Ramirez and Castaneda, 1974]. Chicano children, for example, were found to be more "field dependent", as indicated by their performance on various tests. On the basis of such findings Ramirez [1973] hypothesized that the low academic achievement of Chicano children was primarily due to a mismatch between the instructional style of mainstream education in the United States and the preferred learning style of these children. Attempts were therefore made to design education programs which were considered to be more appropriate for field dependent children.

As Duncan and De Avila [1980] point out, however, an examination of these studies indicates a general failure to control for the language proficiency of these children in either English or Spanish [see also De Avila and Duncan, 1980, for further criticisms of this work]. In addition, recent research by Buriel [1978] fails to support the view that Chicano children are more field dependent than Anglo American children. In fact, research by Duncan and De Avila [1979] shows that when relative linguistic proficiency is taken into account, fully fluent Spanish-English bilinguals are more field *independent* than either monolingual English or monolingual Spanish children. This finding is consistent with work cited earlier indicating that bilingualism can positively affect cognitive growth.

Socio-cultural and attitudinal factors have also been offered as explanations of the lower educational attainment of Chicano children [Tucker, 1977]. Unlike the Canadian French immersion programs, in which the students spoke the dominant language of the culture, a language of higher prestige and ascribed status (i.e., English), the home language of Chicano children (i.e., Spanish) was typically treated as if it were a speech impediment which needed to be corrected as soon as possible by *replacing* it with the dominant language of the culture. Another difference between the Canadian immersion programs and those in the United States was that the Canadian immersion students were normally grouped together in classes in which *none* of the children spoke the language of instruction. In contrast, the language minority children in the United States were mixed together with students who spoke the language of the school. Not surprisingly, the "submerged" language minority children experienced frustration in communicating with teachers and fellow students who were not familiar with their language or culture.

While socio-cultural factors certainly contribute to the poor academic performance of language minority children, it has been argued that specifically linguistic factors also play an important role. In particular, Cummins [1979] has proposed the "threshold" hypothesis which attempts to resolve the apparent inconsistencies in the results of studies of the relationships between bilingualism and cognitive growth. According to this hypothesis, there is a threshold level of bilingualism (or relative linguistic proficiency, to use Duncan and De Avila's terminology) which the child must attain in order to derive potential benefits from bilingualism, and another (lower) level that the child must attain in order to *avoid* possible negative effects on academic achievement. Table 1 summarizes the predicted effects of different language categories (as defined earlier) on academic achievement.

Table 1. Effects of different language categories on academic achievement [after Cummins, 1979]

Language Category	Effect on Academic Achievement
Balanced Bilinguals	Positive
Partial Bilinguals, Monolinguals	Neither Positive nor Negative
Limited Bilinguals, Limited Monolinguals	Negative

As can be seen in this table, negative academic effects are hypothesized to result from low levels of competence in both languages, which Cummins refers to as "semilingualism". Cummins suggests that, unlike Canadian immersion students, many language minority children in the United States come to school with less than native-like command of the vocabulary and syntactic structures of both the first and second languages. Recent research cited by Cummins indicates that this lack of proficiency in both languages appears to result from the child's being exposed to both languages in an unsystematic way prior to coming to school. Consequently, the two languages are not clearly separated, and the child's speech tends to be characterized by code switching (changing from one language to another within the boundaries of sentences), with each language acting as a "crutch" for the other. This would explain the low academic achievement of these children, since reading in either language would be retarded as a result of insufficient knowledge of vocabulary and syntactic rules.

Semilingualism may also account for the low levels of second language learning observed among many language minority children. According to a second hypothesis advanced by Cummins [1979], the "developmental interdependence" hypothesis, the level of competence which a child achieves in a second language is partially a function of the level of competence the child has achieved in the first language at the time when intensive exposure to the second language begins. Recent research on second-language learning [e.g., Mace-Matluck, 1979] indicates that learning a second language is not the same as acquiring language skills "from scratch". Rather, it appears to be a matter of adapting or extending existing skills and knowledge. However, if the child's language skills in his first language are poorly developed, his learning of the second language may be retarded, since he will not be able simply to extend and adapt primary language skills.

Although certain categories of bilingualism appear to have negative effects on academic achievement and second-language learning, two forms are thought to have neither positive nor negative effects (see Table 1), while fully fluent bilingualism is hypothesized to result in *positive* effects on academic achievement. The mechanism through which fully fluent bilingualism exerts its positive effects is the subject of the next section of the paper.

Bilingualism and Metalinguistic Awareness

The central claim of this section is that metalinguistic awareness is the primary variable mediating the positive effects of bilingualism on academic achievement. The argument, in summary form, is that fully fluent bilingualism results in increased metacognitive/metalinguistic abilities which, in turn, facilitate reading acquisition which, in turn, leads to higher levels of academic achievement.

De Avila and Duncan [1979] were among the first to suggest a possible link between fully fluent bilingualism and metacognitive development, the latter referring to the child's developing awareness of how he can *control* his intellectual processes in problem solving tasks [see Flavell, 1977, 1978, 1981]. Duncan and De Avila [1979] hypothesized that "the child who grows up with more than one language, thus acquiring either simultaneously or sequentially two linguistic codes for symbolically manipulating the environment, should enjoy an increased metacognitive awareness" [p.20]. They further hypothesized that "this awareness would be manifested not only on verbal tasks but also in superior information processing and problem solving strategies required on tasks of general intelligence and cognition" [pp.20-21]. In support of these predictions Duncan and De Avila [1979] found that fully fluent Spanish-English bilingual children achieved higher scores than both monolingual English and monolingual Spanish children on measures of intellectual development and cognitive style which required metacognitive processing.

A question that remains unanswered in De Avila and Duncan's account, however, is why children who acquire two languages should "enjoy an increased metacognitive awareness". A possible explanation is provided by Ben-Zeev [1977a], who hypothesizes that "bilinguals become aware of their languages as internally consistent systems more than do other children because this kind of understanding provides a way of separating their languages from each other" [p.45]. This view is consistent with recent research [e.g., Kirsner et al., 1980] which indicates that the two language systems of bilinguals, while perhaps sharing a common supra-lingual system in memory, are functionally independent.

The suggestion, then, is that the process of separating two languages into functionally independent systems results in an increase in "metalinguistic" awareness. Cazden [1976] defines metalinguistic awareness as "the ability to make language forms opaque and attend to them in and for themselves", and suggests that it is a "special kind of language performance, one which makes special cognitive demands, and seems to be less easily and less universally acquired than the language performances of speaking and listening" [p.603]. Perhaps the "special cognitive demands" to which Cazden refers are similar to or identical with what Duncan and De Avila and others call "metacognitive" abilities.

Tunmer and Herriman [this volume] maintain that metacognitive control is a more fundamental ability than metalinguistic awareness, one which is reflected in a wide range of situations and tasks, including those that require reflecting upon and manipulating the structural features of spoken language. Accordingly, they define metalinguistic awareness as the ability to invoke *control* processing to perform mental operations on the products of the mental mechanisms involved in sentence comprehension and production. If metalinguistic performances do indeed reflect a common underlying cognitive capability (i.e., metacognition), then such performances should be correlated with performances on other non-metalinguistic tasks which involve component skills that are metacognitive in character. Evidence in support of this prediction can be found in studies by Hakes et al. [1980] and Tunmer and Fletcher [1981], where positive correlations were reported between measures of metalinguistic awareness and concrete operations, and metalinguistic awareness and conceptual tempo (reflectivity-impulsivity).

These results and the hypothesis advanced by Ben-Zeev [1977a] appear to provide an explanation of the findings reported by Duncan and De Avila. As children in the process of becoming bilingual develop a more analytic orientation to linguistic structures in order to separate the two target languages into functionally independent systems, they automatically acquire higher levels of metacognitive functioning, since cognitive control is necessary to perform metalinguistic operations. In support of this suggestion are several recent studies which seem to indicate that bilinguals do, in fact, enjoy superior metalinguistic, as well as metacognitive, functioning [e.g., Ben-Zeev, 1977a,b; Ianco-Worrall, 1972; Cummins, 1978; Feldman and Shen, 1971]. These studies have attempted to examine systematically the claims of earlier observers that bilingual children have precocious abilities of what are now labelled a "metalinguistic" kind. The studies are based in part on the observations of Leopold [1949] and the theoretical investigations into thought and speech conducted by Vygotsky [1962].

Leopold was one of the first researchers to note the positive effects of bilingualism. He kept a detailed diary record of the observations of his daughter, Hildegard, who was being raised to be bilingual. An important result of her bilingualism which was to be investigated systematically in later studies, was an awareness of the arbitrary nature of language. As Leopold [1961] expressed it, there was "a noticeable looseness of the link between the phonetic word and its meaning" [p.358]. He observed that from a very early age his daughter could tell the same story in two languages and switch vocabulary meaningfully in traditional nursery rhymes. Moreover, she showed an eagerness and interest in immediately learning in her other language the name of any newly-acquired label.

Leopold suggested that both the ability to manipulate words and the interest in language for its own sake seemed to arise from a developing awareness of the loose connection between the word and the object denoted or, as it would now be expressed, a developing awareness of the arbitrary nature of the relationship between the phonological realization of a word and its meaning. The development of these abilities, Leopold further suggested, was a direct outcome of becoming bilingual. Until recently, however, support for this connection between certain metalinguistic abilities and bilingualism relied solely on Leopold's observations of his daughter.

The other influential early work in this area was that of Vygotsky [1962]. His seminal, experimental work carried out in the 1920's and 1930's on the relationship between aspects of thought and language resulted in the development of an experimental technique which has formed the methodological basis of most recent investigations. This technique involved asking children a series of questions about the relationship between words and their referents, such as whether one could call a cow *ink* and ink a *cow*. On the basis of their responses to these questions, Vygotsky concluded that young children initially regarded words as inherent properties of their referent objects. For example, most young children denied that one could call a cow *ink* and ink a *cow*, which suggests they were unable to interchange the names of objects. The justification of their denials supports this view. Children claimed that an animal was called a cow because it had horns. The reason young children were generally unable to perform this task, according to Vygotsky, was because "an interchange of names would mean an exchange of characteristic features, so inseparable is the connection between them in the child's mind" [p.129].

It was important to Vygotsky to investigate how this fusion of word and meaning could be severed, as he felt that it was an essential feature of mature cognitive development that there should be "signification independent of meaning, and meaning independent of reference" [p.130]. Without this separation of word from meaning the child would be unable to control his cognitive processes, or in Vygotsky's terms, he would not be "fully able to formulate his own thought and to understand the speech of others" [p.130]. Using more recent terminology, Vygotsky's work was concerned with the development of metalinguistic and metacognitive abilities, abilities which he regarded as essential to mature linguistic and cognitive functioning, and which Leopold had claimed were an outcome of bilingualism. Largely as a result of Vygotsky's and Leopold's early work, subsequent research on the relationship between metalinguistic awareness and bilingualism has concentrated on bilingual children's awareness of the arbitrary nature of language, as reflected in their ability to separate words from their meanings.

Feldman and Shen [1971], for example, compared the performance of 5-year-old Spanish-English bilinguals and English monolinguals on three tasks, which they regarded as forming a logical sequence of skills. These tasks were object constancy, naming, and the use of names in sentences. The object constancy task was designed to ensure that the children had developed sufficiently in a Piagetian cognitive sense to perform the two language tasks. In the naming task, children were asked to play a "name game" which required them to display three types of labelling ability; the use of common names (e.g., call a cup *cup*); the learning of nonsense names (e.g., call a cup *wug*); and the switching of common names (e.g., call a cup *plate*). In the third task, children were tested on their ability to use the three kinds of labels from the naming task (common, switched common, and nonsense labels) in simple relational sentences, such as "The cup is on the plate". While the performance of monolinguals and bilinguals was equal in relation to knowledge of names and ability to acquire new names, the bilinguals were superior in switching names and using names in sentences. These results would seem to indicate that while bilingualism does not necessarily prove advantageous in all areas of language functioning, it does appear to be advantageous in the more advanced language skills which require flexibility and the separation of the referent from its label.

Another study describing the positive effects of bilingualism on metalinguistic development was Ianco-Worrall's [1972] investigation of nursery school and grade 2-3 Afrikaans-English bilinguals and monolinguals. She adapted Vygotsky's questioning technique to examine whether bilingualism leads to an "earlier realization of the arbitrary nature of the name-object relationship" [p.1390].

In the first of two experiments, she tested the children's semantic or phonetic preferences by asking them to choose words that were either semantically or phonetically related to a stimulus word (e.g., when presented with the words *cap, can* and *hat*, the children were required to choose which of the latter two words was most like *cap*). The second experiment was based on Vygotsky's [1962] work. Children were asked firstly to explain the names of six objects (e.g., "Why is a dog called *dog*"?). They then had to say whether or not the names could be interchanged (e.g., "Could you call a dog *cow* and a cow *dog*?"). In a variation of Vygotsky's original procedure, Ianco-Worrall found it necessary to alter this latter question for the older groups of children. In response to the original form of the question, these children had denied that names could be interchanged, as they felt that the social and linguistic conventions could not be broken. Hence, the question was rephrased in a form that asked whether *in principle* one could interchange the names of things (e.g., "Suppose you were making up names for things, could you call a cow *dog* and a dog *cow*?"). The third part of this second experiment involved an interchange of names (e.g., "Let us play a game. Let us call a dog *cow*". The children were then asked the questions, "Does this *cow* have horns?" and "Does this *cow* give milk?"). [See Ianco-Worrall, 1972, pp.1394-1395].

While Ianco-Worrall found no differences in semantic/phonetic choice amongst the older children, there was a noticeable difference in the younger children's choices. More than half of the younger bilinguals chose words on the semantic dimension, whereas this choice was made by only one of the monolinguals. Ianco-Worrall interpreted these results as support for Leopold's claim that bilinguals developed an earlier separation of word sound from word meaning, since their performance indicated an awareness that a single referent can have more than one phonological realization.

In the second experiment bilinguals and monolinguals did not differ on the first part of the experiment, which concerned the explanation of names. When asked why a cow was called a *cow*, for example, most children would say something like, "because you milk it". However, in the second part of the experiment, which required an interchange of names, there were marked differences between monolinguals and bilinguals in both age groups. The majority of monolinguals denied that the names of objects could be interchanged, while most of the bilinguals agreed that it could be done, in principle. In the third part of the experiment, the children were asked to interchange names in a game setting. In this case, there were no differences between monolinguals and bilinguals in either age group, although performance improved with age.

The findings of this second experiment suggest that the realization that names are arbitrarily assigned to objects is one that develops later than the ability to separate the qualities of an object from its name. Ianco-Worrall summarized her findings by noting that, while the bilinguals and monolinguals did not differ in the way they separated labels from objects in play, both age groups of bilinguals excelled over their monolingual peers when the question of whether names could be interchanged required "the formulated concept that names are arbitrarily assigned to things" [p.1399]. Thus, Ianco-Worrall's work seems to indicate that it is metalinguistic awareness that differentiates the performance of bilinguals from that of monolinguals. This is especially evident on such name interchange tasks as the one she described, which required some "concept" of the arbitrariness of language.

Although the results of both experiments seem to indicate that bilinguals have superior metalinguistic abilities, Ianco-Worrall was cautious in drawing general conclusions from her findings about the effects of bilingualism on cognitive growth, preferring instead to recommend that further research be conducted into the nature of the relationship between bilingualism and cognitive development. At the very least, however, her data support Leopold's claim regarding the early acquisition of certain metalinguistic abilities in bilingual children.

Following Ianco-Worrall's recommendation for more direct research, Ben-Zeev [1977b] conducted a study using Hebrew-English bilingual children and two comparison monolingual groups to explore the hypothesis (referred to earlier in this paper) that mutual interference between the bilingual child's two languages "forces the child to develop particular coping strategies which in some ways accelerate cognitive development" [p.1009]. These coping strategies which Ben-Zeev saw as developing in children in the process of becoming bilingual are ones which require "increased scanning of verbal input" and attention to the linguistic rule system. Bilingual children, it was hypothesized, process syntactic rules "with special flexibility", which arises from their having two referent symbols for many objects, actions, and attributes, leading them to learn at a very early age that words are arbitrary, rather than intrinsic, labels of objects. Ben-Zeev suggested that this metalinguistic insight results in a particular approach to language learning, which she called "hypothesis testing".

To test the hypothesis that bilingual children develop greater cognitive and linguistic flexibility, Ben-Zeev administered a battery of tests to the Hebrew-English bilinguals and monolinguals aged between 5 and 8 years. Children's flexibility of syntactic rule usage was examined in two tests. The first was a verbal transformation test, in which the children were presented repeatedly with the same nonsense word (e.g., *flime*) and asked to say when they thought the stimulus word changed. Previous research had shown that subjects often transformed such stimulus words in response to an illusion that seemed to depend on the operation of a skilled reorganization mechanism which had developed to aid speech perception. As expected, bilinguals reported more verbal transformations at an earlier age than monolinguals, which could be taken to indicate that they had developed a more skilled speech perception mechanism to cope with the demands of interpreting two language systems.

In a second test, symbol substitution, children were required to substitute one meaningful word for another within a fixed sentence frame. This task, which was similar to Ianco-Worrall's name interchange task, was of the following form: "You know that in English this is named *airplane*? In this game its name is *turtle*. Can the *turtle* fly? How does the *turtle* fly?" Children were then asked to substitute words violating an obligatory selection rule of the language: "For this game the way we say *I* is to say, *macaroni*. So how do we say, 'I am warm'?" Bilinguals were expected to show superior performance on this task as it was hypothesized that they would have a more analytic approach to sentence structure.

As expected, bilinguals of both age groups were superior on both parts of the symbol substitution task. Ben-Zeev interpreted these results as indicating the bilinguals' "freedom from word magic" (or the ability to separate the word label from the object) and also their "ability to set aside selectional rules and categorization rules of sentence syntax when this is required" [p.1015]. The bilingual children's generalized readiness to reorganize their perceptions appears to arise from their understanding of the arbitrariness of syntactic structure and the ability to treat words as desemanticized units within a larger syntactic code system. Thus, the general linguistic strategies for coping with bilingualism which seem to be expressed initially in metalinguistic abilities are also evident in more general cognitive strategies, such as those of imputing structure and being ready to make structural reorganizations.

Some further observational evidence of bilingual children's superior metalinguistic abilities arises from Lambert and Tucker's [1972] report of the St. Lambert immersion experiment, which was discussed earlier. Unlike monolinguals, the bilingual children engaged in what Lambert and Tucker called "incipient contrastive linguistics", a process which they described as one of "comparing and contrasting two linguistic systems" [p.207]. This process began in a rudimentary way when the bilinguals engaged in a kind of translation game (i.e., they would give the two names for various objects). They then exhibited a more obvious metalinguistic awareness by, for example, indicating that they realized that "Bonjour, mes enfants" was another way of saying "Good morning, boys and girls". However, the most interesting observation from the study was that the bilingual children then proceeded to make an analysis of the words and structure of each language, just as Ben-Zeev [1977a,b] reported. For example, they seemed to realize that *Bonjour* could be broken down into *bon* and *jour*, which take on similar functions as *good* and *day*. This process of comparison became more systematic as the children then began to notice differences in word order, gender and so on in each language. Moreover, as Leopold had reported, the bilingual children showed a great interest in this process of comparison, which seemed to help them in their linguistic development by providing intrinsic motivation. Finally, Lambert and Tucker noted what they called "linguistic detective skills", which they described as "an attentive, patient, inductive concern with words, meanings and linguistic regularities" [p.208]. As described, these skills seemed to constitute the metalinguistic ability which Ben-Zeev [1977a,b] referred to as an analytic approach to language.

Lambert and Tucker also saw a link between their findings and Vygotsky's work. They suggested that the process of becoming bilingual was analogous to the acquisition of scientific concepts in Vygotsky's theory. Vygotsky [1962] had argued that there were two kinds of concepts -- "spontaneous" or familiar, everyday concepts, and "scientific", or school-learned concepts. Spontaneous concepts in the primary school child are often erroneous, because as Vygotsky expressed it, the child "is not conscious of his own act of thought" [p.108] and is therefore unable to work with these concepts. When he is introduced to scientific concepts, however, he is explicitly made aware of the concepts through verbal definitions and this leads to greater flexibility in handling these scientific concepts. This flexibility is later transferred to spontaneous concepts.

Vygotsky suggested an analogy between concept learning and foreign language learning: "The influence of scientific concepts on the mental development of the child is analogous to the effect of learning a foreign language" [p.109]. The advantages of foreign language learning, Vygotsky outlined as follows:

> In one's native language, the primitive aspects of speech are acquired before the more complex ones. The latter pre-supposes some awareness of phonetic, grammatical, and syntactic forms....a foreign language facilitates mastering the higher forms of the native language. The child learns to see his language as one particular system among many, to view its phenomena under more general categories, and this leads to awareness of his linguistic operations. [p.109]

Although Vygotsky was primarily concerned with foreign language learning in a school setting, his remarks would seem to apply to second language learning in non-school settings as well. Thus, becoming bilingual by whatever means can be viewed as a way of developing the child's metalinguistic awareness with consequent beneficial effects on the child's performance in both languages.

Cummins [1978] attempted to examine more systematically these observations of metalinguistic abilities, and indeed, seems to have been the first to label as "metalinguistic" the ability to comprehend and verbalize the arbitrary nature of the word-referent relationship reported by Leopold, Ben-Zeev and Ianco-Worrall. Using Irish-English bilingual children from grades 3 and 6 (who were selected on fairly stringent linguistic criteria) and monolingual English speakers as a comparison group, he employed tasks developed by Osherson and Markman [1975] to assess the children's ability to examine language in an objective manner and to evaluate nonempirical contradictory and tautological statements.

Objectivity in language was assessed using three tasks. The first concerned the relationship between meaning and referent; specifically, whether a word retains its meaning when its empirical referent is assumed to have been destroyed (e.g., "Does the word *giraffe* have any meaning now that they are all dead?"). The second task tested the child's conception of arbitrariness of language (e.g., "Suppose you were making up names for things, could you call the sun the *moon* and the moon the *sun*?"). The third task explored the child's understanding of the nonphysical nature of words (e.g., "Is the word *book* made of paper?").

To assess their ability to evaluate nonempirical contradications and tautological statements, the children were presented with coloured chips and then a number of statements which they had to judge (in reference to the chips) as either true, false, or impossible to know. The children were also required to justify their responses. Appropriateness of justifications was judged on the children's indication of an awareness of the contradictory or tautological nature of the statements.

The results showed that the bilinguals at both grade levels had greater awareness of the arbitrary nature of the word-referent relationships and were better able to evaluate nonempirical contradictory statements. These results support the observations of Leopold [1961] and Lambert and Tucker [1972], the research findings of Feldman and Shen [1971], Ianco-Worrall [1972] and Ben-Zeev [1977a,b] and the thesis advanced by Vygotsky [1962] that learning another language can promote metalinguistic awareness.

Given the evidence which suggests that fully fluent bilingualism results in superior metacognitive/metalinguistic functioning, a question that remains is how these changes bring about higher levels of academic performance. One possibility that can be considered is that increased metalinguistic abilities facilitate the acquisition of reading skills which, in turn, leads to higher levels of academic performance. As Cummins [1979] argues, "the primary academic task for the child is learning how to extract information efficiently from printed text and subsequent educational progress largely depends upon how well this task is accomplished" [p.237].

Research is accumulating which suggests that the development of metalinguistic abilities may be central to learning to read [see Ehri, 1979; and Tunmer and Bowey, this volume, for reviews]. Most researchers view reading as a derived skill that builds upon spoken language. According to this view the child's fundamental task in learning to read is to discover how to map the printed text onto his existing language, a task which requires the ability to deal explicitly with the structural features of spoken language. To the extent that bilingualism increases the child's metalinguistic abilities, reading achievement and subsequent academic progress should be enhanced. While studies have found a positive relationship between high levels of bilingualism and metalinguistic awareness, and fully fluent bilingualism and academic achievement (including reading), few, if any, have examined the relationship between all three. Metalinguistic awareness may therefore be the primary mechanism through which bilingualism exerts its positive effects on academic functioning, a possibility that awaits further investigation.

Implications for Bilingual Education

As noted earlier, while immersion programs for language *majority* students can result in high levels of functional bilingualism and academic achievement, similar programs for language *minority* children have generally been unsuccessful. These children tended to lag further and further behind, especially in verbal skills. Two other general types of program which have been used with language minority students are "transitional" programs and "maintenance" programs [see Mackey, 1972, 1977].

In standard transitional programs the student's native language is used as the medium of instruction during the period in which the second language is being acquired. For example, if the first language is Spanish and the second language English, the child is initially taught reading and the content areas in Spanish, while at the same time receiving intensive instruction in English as a second language (ESL). As the child moves up the grade levels, instruction in Spanish is gradually decreased as instruction in English is increased. The ultimate aim of transitional programs is to take the child from a dual-language curriculum to a single-language curriculum. These programs are based, in part, on the assumptions that it is pedagogically unwise to attempt to teach a child to read in a language that he does not understand, and that reading skills acquired in the first language readily generalize to the second.

Maintenance programs also utilize both languages as vehicles for teaching and learning. However, unlike transitional programs, after instruction in the first language is gradually decreased and instruction in the second increased, instruction in *both* languages continues in roughly equal amounts as the student moves up the grade levels. The aim of these programs, therefore, is to *maintain* the child's first language so that he can think and function in either language.

In countries in which bilingualism and biliteracy are *not* the goals of bilingual education programs, transitional programs are usually preferred over maintenance programs. This is especially true in the United States, where the primary purpose of bilingual education programs is mastery of the English language. However, in view of the evidence reviewed in the present chapter concerning the positive effects of bilingualism on children's cognitive and academic functioning, it would appear that greater consideration should be given to implementing maintenance programs for language minority students. Moreover, if Cummins' [1979] developmental interdependence hypothesis is correct, initial instruction in the first language would facilitate the second-language learning of children whose language skills in their first language are poorly developed. Although bilingualism and biliteracy would be a consequence of maintenance programs, they would not necessarily be the primary goals. Rather, improved academic functioning and mastery of the English language could still be the primary aims of dual-language programs. The suggestion, however, is that maintenance programs may be a better way of achieving these goals.

Concluding Remarks

To briefly summarize, we have argued in this chapter that, contrary to earlier work, recent research suggests that bilingualism can produce positive effects on cognitive growth and academic achievement. It was suggested that a primary source of the conflicting findings reported earlier in the literature was the failure to distinguish degrees of bilingualism. Accordingly, we introduced De Avila and Duncan's [1979] notion of relative linguistic proficiency and showed how it could be used to develop operational definitions of the different types of bilingualism. These definitions were employed in discussing Cummins' [1979] threshold and interdependence hypotheses, which propose that bilingualism produces positive or negative effects depending on the type of bilingualism involved. We then considered the mechanism through which fully fluent bilingualism exerts its positive effects and hypothesized that metalinguistic awareness acts as the primary intervening variable mediating the positive effects of bilingualism on academic achievement. We concluded the chapter by suggesting that greater consideration be given to implementing bilingual education programs for language minority students which maintain their first language, since this would enable them to enjoy the potential benefits of bilingualism, as well as facilitate their learning of the second language.

3.4 Metalinguistic Awareness and Education

Michael L. Herriman and Marion E. Myhill

In earlier chapters in this volume there has been discussion of the identity, form and genesis of metalinguistic awareness in children. Seen from both the conceptual and empirical perspectives in the current literature on the topic it appears possible to identify metalinguistic awareness in conjunction with a series of changes in cognitive processing which take place during middle childhood. It seems that the ways in which many children view the world and solve problems are qualitatively different from those of the earlier periods of development. The clearest statement of this kind is to be found in Piaget's work where he classifies the "operations" of the child in the concrete operational period as being of a different logical kind--prior to this stage the child's schemes are pre-operational, showing no real indication of a sense of order or "classification". One does not have to accept the Piagetian paradigm, but it is useful to consider the Piagetian view as a framework with which to discuss the claims of other developmentalists [e.g., Cazden, 1975; Donaldson, 1978]. Of particular note is the *qualitatively* different way the child performs tasks and views the world when compared with the earlier period. Flavell [1978, 1980] refers to the development of a set of different cognitive activities which he describes as metacognitive. Similarly, the way in which metalinguistic awareness is identified in the literature preserves its distinctiveness from the language activities which precede it. Hakes et al. [1980], in discussing the kind of linguistic developments "that might parallel" cognitive developments claim that in middle childhood there is "...[an] emergence of a set of new ways of dealing with language, ways that are different from and require cognitive abilities going beyond those involved in understanding and producing utterances" [p.97].

Some writers have suggested that it is not a coincidence that these changes in childhood development take place at the age period most societies have determined is appropriate for the commencement of formal schooling [see for example Donaldson, 1978]. The question of whether there is a cause and effect relationship between these co-occurring events or indeed whether there is any relationship here at all, has been taken up in preceding chapters in this volume. Of importance to education are the questions;

(i) do metalinguistic skills relate significantly to the tasks of education, and in particular, schooling,

(ii) can metalinguistic skills be increased by training, and,

(iii) if so, does training transfer to either specific academic skills or general scholastic achievement?

On the first question it is necessary to emphasize the notion of *significance* for the reason that correlations will not necessarily establish a cause. Metalinguistic skills such as phonemic segmentation ability will predict early reading success, but so will alphabet recognition, yet we would not consider seriously training or teaching alphabet recognition in the belief that it could have any transfer value to reading (except in the most trivial sense). The important point is to establish the centrality (or otherwise) of metalinguistic skills to early educational attainments. In the chapters in this volume by Tunmer and Bowey, and Tunmer and Myhill, the existing evidence was reviewed and arguments were made for the implication of metalinguistic skills in attainment of proficiency in reading and coping with learning a language other than one's native tongue.

On the issue of whether training can increase metalinguistic skills, Bowey [1983], in a study involving a small scale word awareness training to pre-school children (nine lessons of 15 to 20 minutes each), reports "small, but reliable gains in children's concepts of the word as a unit of spoken language....over half had developed relatively sophisticated segmentation strategies which relied on distributional properties of linguistic units" [p.10].

On the issue of transfer, Golinkoff [1978], in her review of studies of the relationship between phonological awareness and reading ability, claims that "if a child has received some type of phonemic awareness training, the literature indicates that the child's reading achievement is likely to be boosted significantly above where it would have been without training" [p.38]. Evidence such as that cited by Rosner [1971], where children who had received training in phonemic awareness skills were seen to profit from early reading instruction when compared with a non-trained group, lends support to this view. As well, Williams [1980] reported that phonological training can facilitate reading instruction. Weaver [1979] has reported that instruction in the use of grammatical structure transferred positively to reading comprehension. Tunmer and Bowey [this volume] have discussed these findings in more detail.

The identification of metalinguistic awareness in tasks such as learning to read, finds more conceptual than empirical support at present. Clearly much more research is needed into issues such as what constitutes the nature of metalinguistic awareness, whether it is a necessary condition for learning to read and whether it can be trained. The present chapter is concerned with the implications of the discussion and reviews conducted above, for educational practice. Its emphasis is on the question of what metalinguistic awareness can contribute to an understanding of early educational or schooling tasks. That this is not a trivial question has been shown by the attention given by educators to Donaldson's [1978] book, *Children's Minds*. Part of Donaldson's view is that we may be underestimating the ability of children to cope with formal schooling, but also that we should look more closely at the under-achieving pupil. She claims that more can be achieved in early education and that we may be overlooking certain abilities that children bring to school. As well, she proposes that a better induction into formal activities such as reading will aid the child in grasping the formal or disembedded character of thinking which schooling demands. In Donaldson's view, reading in particular will contribute to language awareness.

The concern with properly equipping children to cope with early schooling is found also in Russian psychology and education. Karpova [1977] refers specifically to language development at a metalinguistic level when she says;

> ...the preparation of the school child for school instruction cannot be limited to the development of his speech in the process of practical communication and to its enrichment from the viewpoint of vocabulary and grammatical structure. For the successful training of a child, it is absolutely necessary that speech itself as a special reality and the elements of it, particularly the words in the totality of their external (intonational-phonetic) and internal (semantic) aspects, become an object of his consciousness, of his cognitive activity. [p.3.]

The theoretical indebtedness to Vygotsky [1962] and the view of language in its creative potential is quite apparent in both Donaldson's and Karpova's work.

In introducing the concern with education it is difficult to escape what might be seen as the other side of the cause-effect question discussed above; that is, that the idea of schooling might be conceptually identified with the development of just the kinds of abilities said to occur at this stage. It can be asked perhaps whether the identification of the qualitative change taking place in the 5- to 8-year-old is not simply the identification of a schooling effect. However, much of the discussion in the previous chapters supports the view that metalinguistic awareness and schooling effects are conceptually separable and that the main phenomenon, metalinguistic awareness, probably develops in most children independently of the schooling process. This does not deny that many of the tasks of schooling (e.g., certain programs of reading instruction, or attention to certain subskills of reading) will likely increase metalinguistic awareness. Byrne [1981] acknowledges that "most reading programs directly, if not deliberately, teach children about sentences and word structure" [p.86].

To return to the earlier point, it seems that the fact that most societies begin the process of formal schooling in children around the ages of 6 or 7 is indicative of the recognition of developmental changes occurring in language and thinking, changes that is, which are necessary for the tasks of early schooling. This point would be consistent with Donaldson's concerns.

The other relevant aspect of metalinguistic awareness that has been identified in the chapters above (especially in Section II) is its multidimensionality. At least four separate sets of skills or abilities are seen to contribute to general awareness of language form and function -- phonological, lexical, syntactic and pragmatic skills. These categories mark respectively the size of the units said to be separately identifiable as objects of awareness. It is likely that some developmental sequence in the acquisition of awareness occurs as well. It seems plausible that lexical awareness might precede syntactical awareness, and that a realization of the phonemic structure of words requires an awareness of the lexical unit being composed of sub-units which are separately analyzable. Whether an interactive, a hierarchic or simply a sequential relationship is involved is a matter for further research. Another possibility is that the awareness of each category is discrete, although evidence from the study of reading acquisition [Tunmer and Bowey, this volume] suggests that this is unlikely to be the case.

It can be seen that a complex picture emerges here. The multidimensional character of metalinguistic awareness coupled with its suggested relationship to the emergence of more general metacognitive skills, presents both a daunting descriptive task and a wide empirical challenge. But the present concern is to relate this complexity to education, and to see whether any implications for educational practice or activities arise out of the identification of the central phenomenon of concern. We will now examine the picture by looking at metalinguistic awareness in relation to language acquisition, reading and other "secondary" language activities [Mattingly, 1972], and to the more general tasks of schooling.

Language Acquisition

The issue of the relationship of metalinguistic awareness to the *process* of language acquisition is discussed in other chapters in this volume. Most developmentalists accept that language is acquired in a relatively ordered sequence during the first five years [e.g., Brown, 1973; Cazden, 1975; Clark and Clark, 1977; McNeill, 1966], and in the absence of "dramatic handicaps" [Lenneberg, 1968], that proficiency is attained by the greater part of the population despite apparently significant differences and diversities in the linguistic environment to which individual children are exposed. By the time they enter school most children will have attained the basic forms for speaking and understanding speech and further development will likely consist of their learning to use more-syntactically-complex or cognitively- demanding language structures [C. Chomsky, 1969; Palermo and Molfese, 1972; Karmiloff-Smith, 1979a].

Yet it may be that schooling requires more than basic language for success. Some researchers have suggested that children from certain social environments deemed to be disadvantaged, enter school with a language capability which is inadequate to the demands of the situation, in that they are restricted by the kind of language they use. In particular it has been said that these children lack an awareness of the expressive capabilities of language [Bernstein, 1962], or of the cognitively meaningful aspects of language [Hess and Shipman, 1972], or of the structure and intonation of speech [Bereiter and Engelmann, 1966]. This latter quality of language for example, is said to be "necessary for the expression and manipulation of logical relationship" [Bereiter and Engelmann, 1966, p.42]. However, we might ask how problems in a particular area could be related to some general language deficiency. According to Bereiter and Engelmann a significant aspect of "deprived" speech is its lack of any intonational cues in sentence production, such that normal segmental boundaries are not obvious. The sentence "Eh-bih-daw" for "He's a big dog" [quoted by Bereiter and Engelmann, p.34] indicates a failure to stress important distinctive features of the phonemic segments that would seem to be necessary to distinguish, for example, /big/ from /bit/. In earlier chapters it was seen that phonological and lexical skills were significant parts of metalinguistic awareness. The possible identification of language deficiency with low levels of phonological and word-property awareness may therefore provide a means for a more thorough understanding of aspects of standard language development. That is, it could indicate that children exposed to deviant language environments might not develop the kinds of discriminations nor the awareness of the sound properties of language that would seem to be necessary for coping with tasks of early schooling, in particular with reading [Tunmer and Bowey, this volume].

It should be stressed however, that the issue of language deprivation or deviancy is contentious. For example, the work of Labov [1972] has been seen as refuting the earlier views of Bernstein and the various proponents of the language deprivation view. The issue has socio-political as well as racial consequences of a non-trivial kind and it has been avoided on moral grounds as much as it has been dismissed on account of its supposed empirical insignificance. On the evidence currently presented it would seem that the issue cannot be resolved unequivocally. More research is needed, for example, into the phonological and lexical effects of minimal intonational and stress use, and the blurring of distinctive feature boundaries of phonemes; and also into the inferential relationships between syntactic and sentential aspects of language. Far from endorsing the validity of the language deprivation view though, we wish to suggest that an analysis of linguistic knowledge or deprivation and its consequences for logical thought, is to be better informed by considering the separate skills which may be involved in the development of an awareness of language.

Another aspect of language acquisition which some researchers suggest may be related to metalinguistic properties is late-emerging syntactical rules [Hakes et al., 1980]. The ability to deploy and comprehend such structures as object-cleft and passive transformations seems to depend upon the ability to separate the agent-action role from the grammatical subject-verb-object form, that is, the canonical form. According to Karmiloff-Smith [1979a], the latter ability is part of the range of metalinguistic skills which develop in the period in question. Educational practice at the school level traditionally seems to have been based on the assumption that language acquisition is structurally complete on entry to school. If this is not an explicit assumption, the lack of general acceptance of language programs supports an implicit assumption. Yet an examination of certain aspects of development suggests that the process of acquisition is far from complete [Karmiloff-Smith, 1979a]. Although many factors related to knowledge of language function may be involved, we believe that the general role of metalinguistic skills in later grammatical acquisition is an issue requiring further research. Clearly we cannot yet specify a precise role for the teacher in implementing strategies suggested by the claim. Yet it would not be beside the point to suggest that teachers should emphasize such matters as the logical relationships between main and subordinate clauses and the general qualifications implicit in complex grammatical constructions (for example, mood and aspect of verbs for the expression of such logical categories as hypothetical and counterfactual situations). The ability to sustain the thread of an argument or inference in a piece of extended prose is in part dependent upon the ability to follow the variances in grammatical construction in the piece as well as the ability to apply syllogistic reasoning.

Despite the fact that until relatively recently at least, little attention has been given to language acquisition, some educators have recognized a need to cover general kinds of language skills in early childhood education. There are many language development kits and programs available to teachers and language specialists. Interestingly, an examination of a number of these programs reveals that many have incorporated activities that may increase certain aspects of metalinguistic awareness. Thus, in designing programs to aid children's general language development, there has been some realization of the role that metalinguistic awareness may play, although this is not referred to explicitly in programs.

The following examples drawn from recognized programs illustrate activities which may increase certain aspects of metalinguistic awareness. However, it should be noted that the programs do not do so in any systematic or integrated manner.

The *Peabody Language Development Kits* [Dunn et al., 1968] were designed "primarily to stimulate...oral language development", but also "to improve intellectual functioning" and to "enhance future school progress" [p.viii]. There are different kits designed for various levels of development. Level P for example, is for children who are functioning generally at a mental age of 3 to 5 years of age. In the Level P kit there is a greater emphasis on teaching patterns of syntax and simple grammar (e.g., plurals and noun-verb agreement) than in the other kits. The rationale for this emphasis is the assumption that "cognitive development is facilitated by the mastery of the structure of language" [p.viii]. But despite this emphasis on the structural aspects of language development, there are several tasks that *may* help to promote metalinguistic awareness. For example, the skill encouraged in the Listening-Critical Thinking Time task involves aspects of word awareness. Children are given a long word (e.g., *butterfly*) and asked to indicate the short words contained within it (*butter* and *fly*). In another variation of this task children are given the short words (e.g., *cough* and *drop*) and asked for the long word that they make (*coughdrop*). One problem with this activity however is that children may not have an accurate concept of the term *word* [Bowey and Tunmer, this volume], yet the term is included in the instructions. It is therefore quite possible that children would be confused as to the nature and purpose of the task and indeed there is some suggestion in the instructions to teachers that children may not even understand the task at first.

Another task that is similar to some word awareness tasks is the use of a xylophone in some of the "sentence building" activities. In this task each note played on the xylophone represents a separate unit of language, in this case, syllables rather than words. We might ask, however, whether the syllable and its recognition is important to any aspect of language other than intonation or stress patterning. In terms of structure it might be said that the syllable is not an important unit of language at all. In this way some of the tasks in the Peabody kits may work contrary to the goal of metalinguistic understanding. For example, in the Critical Thinking tasks, children are given instructions similar to these: "Today I have some questions to ask you. Some of these questions are silly, but some of them are not silly. If I said 'do fish walk?', you would say, 'No, fish do not walk'. But if I said 'Do fish swim?', you would say, 'Yes, fish do swim'. Now let's try some other questions..."[p.344]. Here it could be argued that the notion of a "silly question" is misleading. Statements or propositions may be "silly" in terms of their semantic content or lack of syntactic well-formedness, but the idea of a grammatical category being silly is wrong, as well as confusing. An important aspect of language development should be the awareness of separation of form and content and, as in the case quoted, syntax and semantics. Here the two are confused.

Another program of interest is *Language and How to Use It* [Monroe, 1970; Schiller et al., 1973] This is a "co-ordinated communications" program designed for pre-school and early primary school children. In the first book of the program, *Beginning Levels*, the aim is to provide a series of language activities for 5- and 6-year-olds which would help these children in the tasks of learning to read and write later on. It provides teachers with suggestions for classroom activities that encourage language development generally. These activities include telling a story associated with a sequence of pictures, and matching and naming colours. There are also some activities that could be said to be concerned with aspects of metalinguistic awareness. These include two activities under the title *That's silly*. In the first activity the concern is with the role of word order and meaning in sentences. Children are shown pictures (e.g., a picture showing a cake on a plate) and the teacher gives them a sentence which is "silly" because the word order is wrong ("The plate is on the cake"). This activity is designed to alert children to the importance of word order to meaning in sentences. The teacher explicitly indicates this to the class by saying: "I'm going to say something silly about this picture. The plate is on the cake! That's silly because the plate isn't really on the cake at all. What should I have said?" [p.111]. A second activity under the *That's Silly* heading is concerned with semantic incongruity. The teacher shows the children a picture (e.g., boys riding bicycles) but says "The boys are riding ponies!" The purpose of this task is to correct the incongruity.

These activities are similar to some of the experimental tasks reviewed by Tunmer and Grieve [this volume]. For example, Gleitman et al. [1972] studied the ability of young children to distinguish between utterances which were well-formed and those which were not (e.g., telegraphic utterances and utterances with reversed word order). In the experiment, children were required to judge utterances as "good" or "silly". Similar studies have also been carried out by de Villiers and de Villiers [1972, 1973]. However, in these research studies, as well as in the language program, children frequently judge sentences as silly on the basis of meaning alone -- they were not always aware of the importance of word order. They concentrated on the content rather than the form of the sentence.

Further examples of metalinguistic awareness in the first book of the program are found in two lessons which are concerned with what we would identify as pragmatic awareness and word awareness. The authors explicitly aim to "awaken young children to the role of language in their lives". They justify this aim by stating:

> By the time children come to school they have mastered the basic system of their language and can give effective expression to thoughts, feelings, ideas and observations consistent with their age and experience. They use language competently, but *without being aware of what it is*. [p. 8, emphasis added]

They suggest that teachers introduce children to "conversation" by asking the children in the group to think of other ways to say the same thing (e.g., "Let's talk", "Let's have a conversation") with the aim of introducing children to "the concept of options, in the expression of an idea". This clearly involves aspects of pragmatic awareness and appreciation of synonymy at the sentence level [as discussed by Hakes et al., 1980].

Another lesson in the task is devoted to an activity involving word awareness. The children in this lesson read a poem called *My Favourite Word* and are then asked to comment on the poem and to talk about various kinds of words -- funny, short, long, etc. However, a problem that arises once again (as with the other language programs) is that these lessons use specialized terminology, such as *conversation, word,* etc. Yet research on word awareness suggests that children at the start of school are generally confused about such terms [Bowey and Tunmer, this volume]. This may detract from the usefulness of such lessons as far as their promoting metalinguistic awareness is concerned.

Talk Reform is a language enrichment program designed by Gahagan and Gahagan [1970] for 5-year-old children who could be described as linguistically disadvantaged in the educational system. The authors believe that language is vital to the acquisition of knowledge and the development of intelligence, but that for many children the school may require a language which is very different from their own. Gahagan and Gahagan, therefore, aim to elaborate and broaden the language spoken by disadvantaged children. Many of the suggestions for language activities in *Talk Reform* seem to be designed to increase metalinguistic awareness, particularly syntactic and pragmatic awareness. The design is apparently unintentional. The authors themselves were concerned with providing activities that would help children who were disadvantaged linguistically due to their exclusive use of the "restricted code" [Bernstein, 1962]. The description of the "restricted code" user provided by Gahagan and Gahagan, however, is one which would probably fit the majority of pre-school children if recent reviews of children's metalinguistic awareness and cognitive development [e.g., Donaldson, 1978] are taken into consideration. Gahagan and Gahagan argue that

> language for a restricted code user is not a medium for exploring logic, or relative truth in argument, nor for weighing opinion. It is tied to its context and in a sense loses its uniquely abstract quality. [p.12]

It could be said that *tying language to a context* is similar to what Donaldson refers to as the *embedded* use of language.

The authors suggest that children who use this code of language exclusively must also learn that "sentences are made of separate words which can be changed and rearranged" [p.14], and that these changes are important for indicating differences in meaning. Some of these activities, then, were designed to encourage language flexibility, but in addition they encourage various kinds of metalinguistic awareness (e.g., word, grammatical and pragmatic awareness).

The authors do come close to explicitly encouraging metalinguistic awareness at one point when they describe how, in the project that formed the basis for the ideas presented in *Talk Reform*, children were given books which were to be used as dictionaries. The aim of this was

> to increase the children's awareness of words in various conceptual and functional hierarchies; to think not only of what a written word denotes or means, but the many attributes of the thing that it means. [p. 58]

The innovation which they suggested may increase awareness of words was that following discussion with the teacher, children should be free to choose the basis of the classification of words in the dictionary. The basis did not have to be the traditional alphabetic one, but could be grammatical function or class concepts (e.g., animals) or another of the children's choice. This task may have the effect of encouraging children to engage in more deliberate reflection on the characteristics of words themselves.

Our brief examination of tasks which have been included in some existing language development programs indicates that, in practical terms, activities that might encourage various aspects of metalinguistic awareness are not necessarily new and may already be in use in some classrooms, although presumably in many cases, without teachers being aware of their potential contribution to metalinguistic awareness.

However, in the existing programs, the tasks which may involve metalinguistic awareness usually form only a small part of the whole program, or relate to only one or two aspects of metalinguistic awareness and are used for purposes other than promoting metalinguistic awareness. Thus although the tasks may contribute to metalinguistic awareness, they are limited in their usefulness unless they can be incorporated in a program that takes into consideration all the components of metalinguistic awareness (phonological, word, grammatical, and pragmatic awareness).

Reading

The programs described above were developed for use in the first years of primary school to promote various aspects of children's general language development. However, if we were to consider the skills thought to be involved in reading acquisition [Tunmer and Bowey, this volume] it would not be surprising to find that many pre-reading or reading readiness activities at the pre-school and early primary grades include tasks that might promote metalinguistic awareness. For example, the Teacher's Manual for the *Target Yellow Phonetic Analysis Kit*, which aims to develop children's skills in "phonetic analysis" as a part of reading acquisition, explicitly mentions the importance of awareness of language in developing these skills;

> Since phonetic analysis is based on somewhat abstract concepts, the development of phonetic analysis skills must be accompanied by the student's increasing awareness of the language that he speaks and hears. [Bamman, 1972,p.7]

Such a comment indicates that a reading approach which incorporates analysis skills is compatible with a concern for developing metalinguistic awareness. In fact, the suggestion above is that both are necessary. Metalinguistic awareness may well be the concept that unifies the various approaches to the teaching of reading so that they can all be involved in the process of learning to read.

The model suggested by Tunmer and Bowey [this volume] indicates that it may be appropriate to use several different teaching methods at various points along the way to fluent reading. According to the model, a language experience approach should be used in the initial stages to familiarise children with the purpose of reading, followed by a "look-and-say" approach to increase children's awareness of words and so on.

Hence, regardless of the approach to the teaching of reading, metalinguistic abilities will be involved in some way in learning to read. Although it is still a theoretical debate as to whether metalinguistic awareness develops as a consequence of learning to read [Donaldson, 1978], or whether the acquisition of reading hinges on the prior development of certain metalinguistic skills, or even whether both reading and metalinguistic awareness are dependent on the development of more general cognitive abilities, we suggest that the development of at least certain aspects of metalinguistic awareness should be tackled directly, that is, by intervention in the early stages of schooling.

There is evidence that young children have difficulty with even elementary aspects of word awareness, and this may make learning to read difficult. Reid [1966] found that many children embarking on the reading process have very little idea of the reading terminology used by teachers (e.g., *word, sentence).* Similarly, Downing [1972] found that children had difficulty with the "abstract technical terms used by teachers in talking about written or spoken language" [p.215]. If the children were explicitly helped to develop awareness of words or a competence with metalinguistic vocabulary for instance, then some of the "cognitive confusion" surrounding initial reading may be avoided [Downing, 1970].

In addition to metalinguistic vocabulary, another form of awareness necessary to reading is that of the conventions of print or text. Clay [1972] has developed tests which use knowledge of conventions of print to diagnose failure in the early stages of reading. As well it has been shown that in first grade, better readers have a better grasp of print related concepts than below average readers [Johns, 1980]. Johns suggests that "teachers may wish to ensure that students are taught the major concepts about print and the language of instruction that are related to reading" [p.547].

General Tasks of Early Schooling

Although the argument has been made for the involvement of metalinguistic factors in language acquisition in school and in the reading process, these aspects of education may not be the only ones related to metalinguistic attainments. The emphasis in metalinguistic awareness upon the ability of the child to control language use suggests that it might be a useful construct for thinking about literacy. In its basic sense literacy is the ability to use language to express and comprehend thought. The literate person is one who has achieved mastery of language to the point where he is expressing his intended meaning as clearly as possible or comprehending the meaning of others fully. There is thus a conceptual link between metalinguistic awareness and language mastery via the notion that both are consciously controlled processes. It was seen previously that the notion of control was also central to the idea of metacognition. Conscious control of the processes involved in producing or extracting meanings is perhaps a mark of literacy. Such a view would be consistent with the view of Vygotsky [1962], who has made a strong claim to the effect that writing is fundamentally linked to conscious control of processes of language and thought:

> Writing...requires deliberate analytical action on the part of the child. In speaking, he is hardly conscious of the sounds he pronounces and quite unconscious of the mental operations he performs. In writing, he must take cognizance of the sound structure of each word, dissect it, and reproduce it in alphabetical symbols, which he must have studied and memorized before. In the same deliberate way, he must put words in a certain sequence to form a sentence. [p.99]

Under this view writing is seen as a separate and different activity from speaking. Its very nature, it would seem, makes it a good medium for the development of an analytic approach to language and thought, through the development of what we would now call metalinguistic awareness. Writing is also important for education in that it is a skill which is necessarily developed through instruction, not spontaneously. Hence just as the development of scientific concepts can be seen to influence the formation of spontaneous concepts [Vygotsky, 1962], writing could be presumed to have an organizing influence on speaking, in that it too could be treated via a more analytical approach, bringing it also under conscious control.

The argument for the connection of literacy to metalinguistic awareness depends in addition upon one's accepting that reading and writing are *not* natural or "primary" activities as are matters such as the acquisition of speech itself [Mattingly, 1972]. This can be realized by a comparison of the relative *uniformity* of speech capability that children from widely differing backgrounds bring to school, with their *wide range* of subsequent performances in reading and writing, taught in the fairly uniform environment of the classroom during the first few years of school. Unlike the effortlessness which speech evidences, the process of writing or reading involves constantly attending to the syntax, semantics and pragmatics of language. The choice of words and grammatical constructions especially in relation to details such as tense, mood and aspects of verbs is crucial to conveying the precise intention of the writer. The requisite abilities seem to have to be learned and practised.

Our argument is that these abilities are developed by some of the activities of early schooling though by no means in the case of every child. To that extent we agree with Donaldson in her emphasis on the effects of early schooling. However, differences in emphasis and techniques within reading instruction can be expected to result in differences in the extent to which children become aware of language and its parts. Some kinds of metalinguistic skills will be practised in certain methods of instruction as well. For example, the phonics method, though discredited as a total approach to teaching reading, will likely make children aware of the phonological structure of segments of language -- even if its method of identifying the sound with its encoding in a set of symbols on paper might at times be arbitrary.

If we allow that reading and writing are not natural activities in the sense described above, it is possible to conceive of literacy as teachable in much the same way as other subjects in a curriculum. Teaching for literacy would mean encouraging students to reflect on their language, to see it as a means for expression that is within their control, and to become able to choose from a range of vocabulary, idiom and style. It is not just a realization that there is a choice, but it is the ability in exercising the choice that is the mark of literacy. We should consider also the possibility that the use of language thus described can reflect back on speech -- to this extent language can *teach* speech, which is the point emphasized earlier in the discussion of Vygotsky's work. The recent emphasis on oral language competence suggests that skills of speech are recognized as important in education.

Our claim then is that metalinguistic awareness is the skill first indicative of a reflective ability in the use of language generally. This view is supported by the studies of Karmiloff-Smith [1979a,b], who equates metalinguistic awareness with other skills which emerge later in childhood (8-11 years), at which stage the child gives evidence of the ability to formulate and make judgements about language. Karmiloff-Smith says that

> from the *consistent* experimental behaviour of the over 8-year-old, we can deduce that by that age the child has attained a more abstract level of linguistic competence, a "metaprocedural level", probably more closely linked to his developing metalinguistic awareness. [1979a, p.314]

Thus described, "metaprocedural" skills can be identified with what have been called "metacognitive skills" [Flavell, 1981], since Karmiloff-Smith [1979a] hypothesizes that "metaprocedural behaviour acts as a control mechanism" in the development of metalinguistic abilities. We have argued that metalinguistic awareness is part of the general cognitive control to which Flavell refers.

The view of literacy suggested here can also be related to that put forward by Olson [1977] in which written language ("text" in Olson's terms) is seen as qualitatively different from speech ("utterance"). Written language involves a specialized activity, requiring deliberate and conscious effort. Of the difference between text and utterance, Olson says

> the bias of written language towards providing definition, making all assumptions and premises explicit and observing the formal rules of logic produces an instrument of considerable power for building an abstract and coherent theory of reality. The development of this explicit, formal system accounts ... for the predominant features of Western culture and for our distinctive ways of using language and our distinctive modes of thought....*Schooling, particularly learning to read is the critical process in the transformation of children's language from utterance to text*. [1977, p.278, emphasis added]

Olson does not say that it is an inevitable property of text that it is of a logically different type to utterance. It is rather that in the actual development of the expository argumentative form, the need to provide an interpretation-free mode of expression was realized.

Though it is never stated quite explicitly, literacy in Olson's view is exemplified in the use of text to express a formal understanding of reality. Since it is in the schooling situation that children mostly confront disciplined studies, which lead to a rapid expansion of their knowledge of the physical and social world surrounding them, their ability to comprehend and discuss this reality would, on the above account, be dependent upon attainment of competence in the use of text. Facility in the use of text seems to be co-extensive with literacy in this view. The dichotomy Olson sets up between utterance and text would also conform to the notions of primary and secondary linguistic activities as put forward by Mattingly [1972]. These ideas can also be linked to the ideas of Vygotsky [1962] as discussed above.

It is not necessary to carry Olson's argument as far as to identify it as necessarily involved in the development of disciplines or "distinctive modes of thought". The emphasis on the purely formal character of text links it sufficiently to the idea of school as the child's introduction to formal or disembedded thought to support the main point of this section -- that metalinguistic awareness is the means whereby the child first attains the level of consciousness of language that can lead to that kind of general awareness of language use which we call literacy. In this claim we do not wish to suggest that there is a stage of metalinguistic awareness and later a stage of literacy, but rather that the terms might represent the ends of a continuum which itself represents one's increasing acquaintance with and awareness of the power of language. As Karmiloff-Smith [1979a] speculates, "it could be that the rich and complex process of language development never ceases entirely" [p.323]. This possibility has been little recognized in educational programs in the past, though there has been an acknowledgement that language must be part of the total curriculum [Britton, 1970].

The notion of an awareness of language is one that can, on the above account, provide a theoretical framework for teaching and learning about language throughout grade school. Language learning would thus involve learning to read, to comprehend the formal language of the classroom, to develop expository, narrative and creative writing skills, and to master where appropriate -- or at least gain a working knowledge of -- other languages used in the community, or foreign languages.

To introduce into all stages of schooling and into the total curriculum the concern with developing awareness of language is to make significant demands on teachers. It would require teachers to be more concerned with their own language use as well as to monitor more carefully the language of all pupils. In this latter task the chief concern will be with the appropriateness of language to express the thought intended. Teachers might also give greater attention to reading development and problems through at least the whole of elementary education, attempting where feasible to look at problems in a broader conception of language deficiency suggested by the metalinguistic view of language skills, especially those of "secondary language activities".

Conclusion

The issues addressed in this paper are drawn from the available literature on the topic of metalinguistic awareness. The connection with education is largely established by extending the conceptual identification of metalinguistic skills with the activities of the classroom, but also by some research which indicates that metalinguistic skills may be necessary for certain educational attainments, especially language related ones.

A survey of some of the literature of, and programs for language-arts and remedial language teaching, showed that there was some recognition of metalinguistic aspects of language understanding in oral language contexts as well as in reading. No program has been developed in which metalinguistic awareness is given a wider role, relating it to the formal tasks of schooling or to the general tasks of language learning for the attainment of literacy. Clearly more research is needed to establish the empirical basis for relating metalinguistic awareness and education and to answer questions about training and transfer of the skills thought to comprise the awareness.

Several potentially informative lines of research are suggested. One would be a large scale training study of matched groups in which one group would be given a program of metalinguistic awareness training and the other a program of similar duration and instructional method but other content. The groups would be later compared with respect to language attainments and academic success. Another study might be a predictive one, where the basis for prediction would be the tested metalinguistic abilities of school beginners. If metalinguistic abilities were related to success in early schooling, then of those children scoring well on tests of metalinguistic abilities, it might be predicted that success in reading and other formal tasks (e.g., mathematics) will follow. Another fruitful test of the general hypothesis might come from a cross-sectional study where matched groups at different age-grade levels might be compared. Nesdale and Tunmer [this volume] review more of the methodological questions associated with research on this topic.

At a general level of investigation more research is needed into the relationship between metalinguistic awareness and metacognition. The work of Hakes et al. [1980] suggests further lines for research of this kind.

A further promising direction of research has been discussed by Karmiloff-Smith [1979a,b] and referred to above. Her concern with "metaprocedural behaviour" of children up to late childhood has extended the chronology of metalinguistic awareness. On the issue of pragmatic awareness, research reported for example by Robinson and Robinson [1980] indicates the importance of understanding the development of communication processes in children. There has been little research into aspects of early schooling to do with communicative skills, however, and the Robinsons' research points to a clear gap in knowledge there. Further discussion of this issue can be found in Pratt and Nesdale [this volume]. The connection between pragmatic awareness and the kinds of logical skills involved in making inferences from related sentences or propositions also requires further exploration.

As well, more research is needed in the area of sub-component metalinguistic abilities; that is, in the further delineation of the skills comprising metalinguistic awareness. Two approaches are required; one pursuing the methodological issues of test design, and the other analyzing the conceptual and empirical adequacy of present tests for identifying metalinguistic skills. With the present emphasis on action research in education, there is wide scope for participation of schools in both research and program evaluation in metalinguistic awareness.

Perhaps Karpova [1977] best summarizes the educational importance of second-order language skills when she says, "A child's realization of speech and its elements is necessary not only for the teaching of writing and reading, but also to enable the child to make the system of knowledge presented to him an object of his study activity" [p.3].

References

ACKERMAN, B. P. Children's use of contextual expectations to detect and resolve comprehension failures. *Journal of Experimental Child Psychology*, 1982, *33*, 63-73.

ALFORD, J. Contextual and lexical effects on reading time. Unpublished doctoral dissertation, The University of Texas at Austin, 1980.

ANDERSEN, E.S. Learning to speak with style: A study of the sociolinguistic skills of children. Unpublished doctoral dissertation, Stanford University, 1977.

ARGYLE, M. *Social interaction.* London: Methuen, 1969.

ARIES, P. *Centuries of childhood.* London: Cape, 1962.

ARLIN, P. Piagetian tasks as predictors of reading and math readiness in grades K-1. *Journal of Educational Psychology*, 1981, *73*, 712-721.

ASHER, S. R. Children's ability to appraise their own and another person's communication performance. *Developmental Psychology*, 1976, *12*, 23-32.

AUCKERMAN, R. C. *Approaches to beginning reading.* New York: Wiley, 1971.

BAIN, B. C., & YU, A. Toward an integration of Piaget and Vygotsky: A cross-cultural replication (France, Germany, Canada) concerning cognitive consequences of bilinguality. In M. Paradis (Ed.), *Aspects of bilingualism.* Columbia, S. C.: Hornbeam Press, 1978.

BAMMAN, H. *Target yellow kit teacher's manual.* California: Addison-Wesley, 1972.

BATES, E. *Language and context: The acquisition of pragmatics.* New York: Academic Press, 1976.

BEAL, C. R., & FLAVELL, J. H. The effect of increasing the salience of message ambiguities on kindergarteners' evaluations of communicative success and message adequacy. *Developmental Psychology*, 1982, *18*, 43-48.

BEARISON, D. Role of measurement operations in the acquisition of conservation. *Developmental Psychology*, 1969, *1*, 653-660.

BEARISON, D. J., & LEVEY, L. M. Children's comprehension of referential communication: Decoding ambiguous messages. *Child Development*, 1977, *48*, 716-720.

BEILIN, H., LUST, B., SACK, H., & NATT, M. *Studies in the cognitive basis of language development.* New York: Academic Press, 1975.

BEN-ZEEV, S. Mechanisms by which childhood bilingualism affects understanding of language and cognitive structures. In P. A. Hornby (Ed.), *Bilingualism: Psychological, social, and educational implications.* New York: Academic Press, 1977. (a)

BEN-ZEEV, S. The influence of bilingualism on cognitive strategy and cognitive development. *Child Development*, 1977, *48*, 1009-1018. (b)

BEREITER, C., & ENGELMANN, S. *Teaching disadvantaged children in the preschool*. Englewood Cliffs, N.J.: Prentice Hall, 1966.

BERKO, J. The child's learning of English morphology. *Word*, 1958, *14*, 150-177.

BERNSTEIN, B. Social class, linguistic codes and grammatical elements. *Language and Speech*, 1962, *5*, 221-240.

BERTHOUD-PAPANDROPOULOU, I. An experimental study of children's ideas about language. In A. Sinclair, R. J. Jarvella & W.J.M. Levelt (Eds.), *The child's conception of language*. Berlin: Springer-Verlag, 1978.

BEVER, T. G. The cognitive basis for linguistic structures. In J. Hayes (Ed.), *Cognition and the development of language*. New York: Wiley, 1970.

BICKLEY, A. C., ELLINGTON, B. J., & BICKLEY, R. T. The cloze procedure: A conspectus. *Journal of Reading Behaviour*, 1970, *2*, 232-249.

BIEMILLER, A. J. The development of the use of graphic and contextual information as children learn to read. *Reading Research Quarterly*, 1970, *6*, 75-96.

BLOOMFIELD, L. *Language*. New York: Holt, 1946.

BOHANNON, J. N. The relationship between syntax discrimination and sentence imitation in children. *Child Development*, 1975, *46*, 441-451.

BOHANNON, J. N. Normal and scrambled grammar in discrimination, imitation and comprehension. *Child Development, 1976, 47*, 669-681.

BOHN, W. E. First steps in verbal expression. *Pedagogical Seminary*, 1914, 21, 578-595.

BOND, B. L., & DYKSTRA, R. The co-operative research program in first-grade reading instruction. *Reading Research Quarterly*, 1967, *2*, 5-142.

BORING, E.G. A history of introspection. *Psychological Bulletin*, 1953, *50*, 169-187.

BOWES, J. Children's play and awareness of language: Some possible relationships to reading. *Working Papers in Language and Linguistics*, 1979, *9*, 42-61.

BOWEY, J. A. Syntactic control in relation to children's oral reading performance. Unpublished manuscript, University of Melbourne, 1982.

BOWEY, J. A. Teaching word awareness to preschool children within a kindergarten setting. In T. G. Cross & L. M. Riach (Eds.), *Issues and research in child development: Proceedings of the second national conference on child development*. Melbourne: Australian Council for Educational Research, 1983.

BOWEY, J. A., TUNMER, W. E., & PRATT, C. The development of children's understanding of the metalinguistic term *word*. *Journal of Educational Psychology*, in press.

BRAINERD, C.J. Judgments and explanations as criteria for the presence of cognitive structures. *Psychological Bulletin*, 1973, *79*, 172-179.

BRAINERD, C. J. *Piaget's theory of intelligence*. Englewood Cliffs, N. J.: Prentice-Hall, 1978.

BRANSFORD, J. D., & JOHNSON, M. K. Contextual prerequisites for understanding: Some investigations of comprehension and recall. *Journal of Verbal Learning and Verbal Behavior*, 1972, *11*, 717-726.

BRANSFORD, J. D., & JOHNSON, M. K. Considerations of some problems of comprehension. In W.G. Chase (Ed.), *Visual information processing*. New York: Academic Press, 1973.

BREKKE, B., WILLIAMS, J., & HARLOW, S. Conservation and reading readiness. *Journal of Genetic Psychology*, 1973, *123*, 133-138.

BRITTON, J. *Language and Learning*. Harmondsworth: Penguin, 1970.

BRODZINSKY, D. M. Children's comprehension and appreciation of verbal jokes in relation to conceptual tempo. *Child Development*, 1977, *48*, 960-967.

BRODZINSKY, D. M., FEUER, V., & OWENS, V. Detection of linguistic ambiguity by reflective, impulsive, fast-accurate and slow-inaccurate children. *Journal of Educational Psychology*, 1977, *69*, 237-243.

BROWN, A. L. Knowing when, where, and how to remember: A problem of metacognition. In R. Glaser (Ed.), *Advances in instructional psychology*. Hillsdale, N. J.: Lawrence Erlbaum Associates, 1978.

BROWN, A. L. Metacognitive development and reading. In R. J. Spiro, B. C. Bruce & W. F. Brewer (Eds.), *Theoretical issues in reading comprehension*. Hillsdale, N.J.: Lawrence Erlbaum Associates, 1980.

BROWN, A. L., & DeLOACHE, J. S. Skills, plans and self-regulation. In R. Siegler (Ed.), *Children's thinking: What develops?* Hillsdale, N. J.: Lawrence Erlbaum Associates, 1978.

BROWN, A. L., & SMILEY, S.S. Rating the importance of structural units of prose passages: A problem of metacognitive development. *Child Development*, 1977, *48*, 1-8.

BROWN, R. *A first language: The early stages*. Cambridge, Mass: Harvard University Press, 1973.

BROWN, R., & BELLUGI, U. Three processes in the child's acquisition of syntax. *Harvard Educational Review*, 1964, *34*, 133-151.

BRUCE, D. J. The analysis of word sounds by young children. *British Journal of Educational Psychology*, 1964, *34*, 158-170.

BRUNER, J. S. *The relevance of education*. London: Allen & Unwin, 1971.

BRUNER, J. S. The role of dialogue in language acquisition. In A. Sinclair, R. J. Jarvella, & W. J. M. Levelt (Eds.), *The child's conception of language*. Berlin: Springer-Verlag, 1978.

BRYANT, P. E. *Perception and understanding in young children: An experimental approach*. London: Methuen, 1974.

BRYANT, P. E., & KOPYTYNSKA, H. Spontaneous measurement by young children. *Nature,* 1976, *260*, 772.

BRYANT, P. E., & TRABASSO, T. Transitive inferences and memory in young children. *Nature*, 1971, *232*, 456-458.

BURIEL, R. Relationship of three field-dependence measures to the reading and math achievement of Anglo American and Mexican American children. *Journal of Educational Psychology*, 1978, *70*, 167-174.

BYRNE, E. B. Reading disability, linguistic access and short-term memory: Comments prompted by Jorm's review of developmental dyslexia. *Australian Journal of Psychology*, 1981, *33*, 83-95.

CALFEE, R. C. Memory and cognitive skills in reading acquisition. In D. Duane & M. Rawson (Eds.), *Reading perception and language*. Baltimore: York Press, 1975.

CALFEE, R. C. Assessment of independent reading skills: Basic research and practical applications. In A. S. Reber & D. L. Scarborough (Eds.), *Toward a psychology of reading*. Hillsdale, N. J.: Lawrence Erlbaum, 1977.

CALFEE, R. C., ARNOLD, R., & DRUM, P.A. A review of *The psychology of reading* by E. Gibson & H. Levin. *Proceedings of the National Academy of Education*, 1976, *3*, 1-80.

CALFEE, R. C., CHAPMAN, R., & VENEZKY, R. How a child needs to think to learn to read. In L. Gregg (Ed.), *Cognition in learning and memory*. New York: Halsted Press, 1972.

CALFEE, R. C., & DRUM, P. A. Learning to read: Theory, research and practice. *Curriculum Inquiry*, 1978, *8*, 183-249.

CALFEE, R. C., LINDAMOOD, P., & LINDAMOOD, C. Acoustic-phonetic skills and reading--kindergarten through twelfth grade. *Journal of Educational Psychology*, 1973, *64*, 293-298.

CANNELL, C.F., OKSENBERG, L., & CONVERSE, J.M. Striving for response accuracy: Experiments in new interviewing techniques. *Journal of Marketing Research*, 1977, *14*, 307-315.

CAREY, S. Less never means more. In R. N. Campbell & P. T. Smith (Eds.), *Recent advances in the psychology of language. Language development and mother-child interaction* (Vol. 1). London: Plenum Press, 1978.

CARR, D. The development of young children's capacity to judge anomalous sentences. *Journal of Child Language*, 1979, *6*, 227-241.

CAZDEN, C. B. Play with language and metalinguistic awareness: One dimension of language experience. In C. B. Winsor (Ed.), *Dimensions of language experience*. New York: Agathon Pre.s, 1975.

CAZDEN, C. B. Play with language and metalinguistic awareness: One dimension of language experience. In J. S. Bruner, A. Jolly & K. Sylva (Eds.), *Play*. New York: Penguin, 1976.

CAVANAUGH, J.C., & PERLMUTTER, M. Metamemory: A critical examination. *Child Development*, 1982, *53*, 11-28.

CHALL, J. S. *Learning to read: the great debate*. New York: McGraw-Hill, 1967.

CHOMSKY, C. *The acquisition of syntax in children from 5 to 10*. Cambridge, Mass.: M.I.T. Press, 1969.

CHOMSKY, N. *Aspects of the theory of syntax*. Cambridge, Mass.: M.I.T. Press, 1965.

CHOMSKY, N. *Language and mind*. (Enlarged edition). New York: Harcourt Brace Jovanovich, 1972.

CHOMSKY, N., and HALLE, M. *The sound pattern of English*. New York: Harper & Row, 1968.

CLARK E. V. Awareness of language: Some evidence from what children say and do. In A. Sinclair, R. J. Jarvella, & W.J.M. Levelt (Eds.), *The child's conception of language*. Berlin: Springer-Verlag, 1978.

CLARK, E. V., & ANDERSEN, E. S. Spontaneous repairs: Awareness in the process of acquiring language. Paper presented at Symposium on Reflections on Metacognition, Society for Research in Child Development, San Francisco, March, 1979.

CLARK, H. H., & CLARK, E. V. *Psychology and language: An introduction to psycholinguistics*. New York: Harcourt Brace Jovanovich, 1977.

CLAY, M. E. *The early detection of reading difficulties: A diagnostic survey*. Auckland, New Zealand: Heinemann, 1972.

COHEN, A. D. The Culver City Spanish immersion program: The first two years. *The Modern Language Journal*, 1974, *58*, 95-103.

COLE, M., & SCRIBNER, S. *Culture and thought: A psychological introduction*. New York: John Wiley & Sons, 1974.

COSGROVE, J. M., & PATTERSON, C. J. Plans and the development of listener skills. *Developmental Psychology*, 1977, *13*, 557-564.

COSGROVE, J. M., & PATTERSON, C. J. Generalization of training for children's listening skills. *Child Development*, 1978, *49*, 513-516.

COWAN, P. A., WEBER, J., HODDINOTT, B. A., & KLEIN, J. Mean length of spoken responses as a function of stimulus, experimenter and subject. *Child Development*, 1967, *38*, 191-203.

COX, M. The effect of conservation ability on reading competency. *The Reading Teacher*, 1976, *30*, 251-258.

CROMER, W. The difference model: A new explanation for some reading difficulties. *Journal of Educational Psychology*, 1970, *61*, 471-483.

CRONBACH, L. J. *Educational psychology*. New York: Harcourt Brace Jovanovich, 1977.

CUMMINS, J. Bilingualism and the development of metalinguistic awareness. *Journal of Cross-cultural Psychology*, 1978, *9*, 131-149.

CUMMINS, J. Linguistic interdependence and the educational development of bilingual children. *Review of Educational Research*, 1979, *49*, 222-251.

CUMMINS, J., & GULUTSAN, M. Some effects of bilingualism on cognitive functioning. In S. Carey (Ed.), *Bilingualism, biculturalism and education*. Edmonton, Alberta: The University of Alberta Press, 1974.

DALE, P. S. *Language acquisition: Structure and function*. Second Edition. New York: Holt, Rinehart & Winston, 1976.

De AVILA, E. A., & DUNCAN, S. E. *Language Assessment Scales, Level I*. San Rafael, Ca.: Linguistic Group, 1977.

De AVILA, E. A., & DUNCAN, S. E. Definition and measurement: The east and west of bilingualism. Unpublished paper prepared for the California State Department of Education, 1978.

De AVILA, E. A., & DUNCAN, S. E. Bilingualism and the metaset. *Journal of the National Association for Bilingual Education*, 1979, *3*, 1-20.

De AVILA, E. A., & DUNCAN, S. E. Field dependence/independence of traditional and dualistic Chicano communities and Anglo communities. In D. Dominguez (Ed.), *Cross-cultural investigations of cognitive style*. Austin, Tx.: Southwest Educational Development Laboratory, 1980.

deMAUSE, L. The evolution of childhood. In L. deMause (Ed.), *The history of childhood*. New York: Harper & Row, 1974.

DENNER, R. Representational and syntactic competence of problem readers. *Child Development*, 1970, *41*, 881-887.

DENNY, D. R. Relationship of three cognitive style dimensions to elementary reading abilities. *Journal of Educational Psychology*, 1974, *66*, 702-709.

deVILLIERS, J. G., & deVILLIERS, P. A. Early judgments of semantic and syntactic acceptability by children. *Journal of Psycholinguistic Research*, 1972, *1*, 229-310.

deVILLIERS, J. G., & deVILLIERS, P. A. Development of the use of word order in comprehension. *Journal of Psycholinguistic Research*, 1973, *2*, 331-341.

deVILLIERS, J. G., & deVILLIERS, P. A. Competence and performance in child language: Are children really competent to judge? *Journal of Child Language*, 1974, *1*, 11-22.

deVILLIERS, J. G., & deVILLIERS, P. A. *Language acquisition*. Cambridge, Mass.: Harvard University Press, 1978.

DICKSON, W. P. (Ed.), *Children's oral communication skills*. New York: Academic Press, 1981.

DICKSON, W. P. Referential communication activities in research and in the curriculum: A meta-analysis. In W. P. Dickson (Ed.), *Children's oral communication skills*. New York: Academic Press, 1981.

DOCKRELL, J., NEILSON, I., & CAMPBELL, R. Conservation accidents revisited. *International Journal of Behavioural Development*, 1980, *3*, 423-439.

DONALDSON, M. Development of conceptualization. In V. Hamilton & M. Vernon (Eds.), *Development of cognitive processes*. London: Academic Press, 1976.

DONALDSON, M. *Children's minds*. Glasgow: Collins, 1978.

DONALDSON, M. Conservation: What is the question? *British Journal of Psychology*, 1982, *73*, 199-207.

DONALDSON, M., GRIEVE, R., & PRATT, C. (Eds.), *Early childhood development and education: Readings in psychology*. Oxford: Basil Blackwell, 1983.

DONALDSON, M., & McGARRIGLE, J. Some clues to the nature of semantic development. *Journal of Child Language*, 1974, *1*, 185-194.

DOOLING, D. J., & LACHMAN, R. Effects of comprehension on retention of prose. *Journal of Experimental Psychology*, 1971, *88*, 216-222.

DOWNING, J. How children think about reading. *The Reading Teacher*, 1969, *23*, 217-230.

DOWNING, J. Children's concepts of language in learning to read. *Educational Research*, 1970, *12*, 106-112.

DOWNING, J. Children's developing concepts of spoken and written language. *Journal of Reading Behaviour*, 1971-1972, 4, 1-19.

DOWNING, J. How children think about reading. In A. Melnick & J. Merritt (Eds.), *The reading teacher*. London: University of London Press, 1972.

DOWNING, J. *Comparative reading*. New York: Macmillan, 1973.

DOWNING, J. *Reading and reasoning*. Edinburgh: Chambers, 1979.

DOWNING, J., & LEONG, C. *Psychology of reading*. New York: Macmillan, 1982.

DOWNING, J., & OLIVER, P. The child's concept of a word. *Reading Research Quarterly*, 1974, *9*, 568-582.

DUNCAN, S. E., & De AVILA, E. A. Bilingualism and cognition: Some recent findings. *Journal of the National Association for Bilingual Education*, 1979, *4*, 15-50.

DUNCAN, S. E., & De AVILA, E. A. Relative linguistic proficiency and field dependence/independence: Some findings on linguistic heterogeneity and cognitive style of bilingual children. In D. Dominguez (Ed.), *Cross-cultural investigations of cognitive style.* Austin, Tx.: Southwest Educational Development Laboratory, 1980.

DUNN, L. M., HORTON, K. B., & SMITH, J. O. *Peabody language development kits: Manual for level P.* Minnesota: American Guidance Service, 1968.

EHRI, L. C. Word consciousness in readers and pre-readers. *Journal of Educational Psychology*, 1975, *67*, 204-212.

EHRI, L. C. Linguistic insight: Threshold of reading acquisition. In T. G. Waller & G. E. Mackinnon (Eds.), *Reading research· Advances in theory and practice.* New York: Harcourt Brace Jovanovich, 1979.

EHRI, L. C., & ROBERTS, K. T. Do beginners learn printed words better in contexts or in isolation? *Child Development*, 1979, *50*, 675-685.

EHRI, L. C., & WILCE, L. C. The influence of orthography on readers' conceptualization of the phonemic structure of words. *Applied Psycholinguistics*, 1980, *1*, 371-385.

EIMAS, P. D. Speech perception in early infancy. In L. B. Cohen & P. Salapatek (Eds.), *Infant perception: From sensation to cognition.* (Vol. 2). New York: Academic Press, 1975.

EIMAS, P. D., SIQUELAND, E. R., JUSCZYK, P., & VIGORITO, T. Speech perception in infants. *Science*, 1971, *171*, 303-306.

ELKONIN, D. B. U.S.S.R. In J. Downing (Ed.), *Comparative reading.* New York: Macmillan, 1973.

ELLIOT, A. J. *Child language.* Cambridge: Cambridge University Press, 1981.

ESON, M. E., & WALMSLEY, S. A. Promoting cognitive and psycholinguistic development. In M. Johnson (Ed.), *Toward adolescence: The middle school years.* Chicago: University of Chicago Press, 1980.

EVANS, M., TAYLOR, N., & BLUM, I. Children's written language awareness and its relation to reading acquisition. *Journal of Reading Behaviour*, 1979, *11*, 7-19.

FELDMAN, C., & SHEN, M. Some language-related cognitive advantages of bilingual five-year-olds. *The Journal of Genetic Psychology*, 1971, *118*, 235-244.

FINN, G. P. T. Ask a silly question: But get a serious answer. Unpublished paper, University of St. Andrews, 1976.

FLAVELL, J. H. *The developmental psychology of Jean Piaget.* Princeton, N. J.: Van Nostrand, 1963.

FLAVELL, J. H. *Cognitive development.* Englewood Cliffs, N. J.: Prentice-Hall, 1977.

FLAVELL, J. H. Metacognitive development. In J. M. Scandura & C. J. Brainerd (Eds.), *Structural process models of complex human behaviour.* Alphen an den Rijn, The Netherlands: Sijthoff and Noordhoff, 1978.

FLAVELL, J. H. Cognitive monitoring. In W. P. Dickson (Ed.), *Children's oral communication skills.* New York: Academic Press, 1981.

FLAVELL, J. H., SPEER, J. R., GREEN, F. L., & AUGUST, D. L. The development of comprehension monitoring and knowledge about communication. *Monographs of the Society for Research in Child Development*, 1981, Serial No.192, *46(5)*, pp.1-55.

FLAVELL, J. H., & WELLMAN, H. M. Metamemory. In R. V. Kail Jr. & J. W. Hagen (Eds.), *Perspectives on the development of memory and cognition.* Hillsdale, N. J.: Lawrence Erlbaum Associates, 1977.

FOPPA, K. Language acquisition - a human ethological problem? *Social Science Information*, 1978, *17*, 93-105.

FORSTER, K. I. Priming and the effects of sentence and lexical contexts on naming time: Evidence for autonomous lexical processing. *Quarterly Journal of Experimental Psychology*, 1981, *33A*, 465-495.

FOSS, D. J., & HAKES, D. T. *Psycholinguistics: An introduction to the psychology of language*. Englewood Cliffs, N. J.: Prentice-Hall, 1978.

FOWLES, B., & GLANZ, M. E. Competence and talent in verbal riddle comprehension. *Journal of Child Language*, 1977, *4*, 433-452

FOX, B., & ROUTH, D. K. Analyzing spoken language into words, syllables and phonemes: A developmental study. *Journal of Psycholinguistic Research*, 1975, *4*, 331-342.

FOX, B., & ROUTH, D. K. Phonemic analysis and synthesis as word-attack skills. *Journal of Educational Psychology*, 1976, *68*, 70-74.

FRANCIS, H. Children's experience of reading and notions of units in language. *British Journal of Educational Psychology*, 1973, *43*, 17-23.

FRASER, C. Communication in interaction. In H. Tajfel, & C. Fraser (Eds.), *Introducing social psychology*. Harmondsworth: Penguin, 1978.

FRIES, C. G. *Linguistics and reading*. New York: Holt, Rinehart and Winston, 1963.

GAHAGAN, D. M., & GAHAGAN, G. A. *Talk reform*. London: Routledge & Kegan Paul, 1970.

GARRETT, M. F. The analysis of sentence production. In G. H. Bower (Ed.), *The psychology of learning and motivation*, Vol.9. New York: Academic Press, 1975.

GELB, I. J. *A study of writing: The foundation of grammatology*. (Revised edition). Chicago: University of Chicago Press, 1963.

GELMAN, R. Cognitive development. *Annual Review of Psychology*, 1978, *29*, 297-332.

GELMAN, R. Assessing one-to-one correspondence: Still another paper about conservation. *British Journal of Psychology*, 1982, *73*, 209-220.

GELMAN, R., & GALLISTEL, C. R. *The child's understanding of number*. Cambridge, Mass.: Harvard University Press, 1978.

GELMAN, R., & SHATZ, M. Appropriate speech adjustments: The operation of conversational constraints on talk to two-year-olds. In M. Lewis & L. Rosenblum (Eds.), *Interaction, conversation, and the development of language*. New York: Wiley, 1977.

GIBSON, E. J., & LEVIN, H. *The psychology of reading*. Cambridge, Mass.: M.I.T. Press, 1975.

GLEASON, J. B. Code switching in children's language. In T. E. Moore (Ed.), *Cognitive development and the acquisition of language*. New York: Academic Press, 1973.

GLEITMAN, L. Metalinguistics is not kid-stuff. Paper presented at the International Reading Research Seminar on Linguistic Awareness and Learning to Read, Victoria, B.C., Canada, June, 1979.

GLEITMAN, L. R., & GLEITMAN, H. *Phrase and paraphrase: Some innovative uses of language*. New York: Morton, 1970.

GLEITMAN, L. R., GLEITMAN, H., & SHIPLEY, E. F. The emergence of the child as grammarian. *Cognition*, 1972, *1*, 137-164.

GLEITMAN, L. R., & ROZIN, P. Teaching reading by use of a syllabary. *Reading Research Quarterly*, 1973, *8*, 447-483.

GLEITMAN, L. R., & ROZIN, P. The structure and acquisition of reading I: Relations between orthographics and the structure of language. In A. S. Reber & D. L. Scarborough (Eds.), *Toward a psychology of reading*. Hillsdale, N. J.: Lawrence Erlbaum Associates, 1977.

GLUCKSBERG, S., & KRAUSS, R. M. What do people say after they have learned to talk? Studies of the development of referential communication. *Merrill-Palmer Quarterly*, 1967, *13*, 309-316.

GLUCKSBERG, S., KRAUSS, R. M., & HIGGINS, T. The development of communication skills in children. In F. Horowitz (Ed.), *Review of child development research* (Vol.4). Chicago: University of Chicago Press, 1975.

GLUCKSBERG, S., KRAUSS, R. M., & WEISBERG, R. Referential communication in nursery school children: Method and some preliminary findings. *Journal of Experimental Child Psychology*, 1966, *3*, 333-342.

GOLDSTEIN, D. M. Cognitive-linguistic functioning and learning to read in pre-schoolers. *Journal of Educational Psychology*, 1976, *68*, 680-688.

GOLDSTEIN, D. Comprehension of linguistic ambiguity and development of classification. *Perceptual and Motor Skills*, 1976, *43*, 1051-1058.

GOLINKOFF, R. M. A comparison of reading comprehension processes in good and poor comprehenders. *Reading Research Quarterly*, 1975-1976, *4*, 623-659.

GOLINKOFF, R. M. Critique: Phonemic awareness skills and reading achievement. In F. B. Murray & J. J. Pikulski (Eds.), *The acquisition of reading: Cognitive, linguistic and perceptual prerequisites*. Baltimore: University Park Press, 1978.

GOODMAN, K. S. Reading: A psycholinguistic guessing game. *Journal of the Reading Specialist*, 1967, *6*, 126-135.

GOODMAN, K. S. The psycholinguistic nature of the reading process. In K. S. Goodman (Ed.), *The psycholinguistic nature of the reading process*. Detroit: Wayne State University Press, 1968.

GOODMAN, K. S. The 13th easy way to make learning to read difficult: A reaction to Gleitman and Rozin. *Reading Research Quarterly*, 1973, *8*, 484-493.

GOUGH, P. B. One second of reading. In J. F. Kavanagh & I. G. Mattingly (Eds.), *Language by ear and by eye*. Cambridge, Mass.: M.I.T. Press, 1972.

GOUGH, P. B. The structure of language. In D. D. Duane & M. B. Rawson (Eds.), *Reading, perception and language*. Baltimore: York Press, 1975.

GOUGH, P. B., ALFORD, J., & HOLLEY-WILCOX, P. Words and contexts. In M. L. Kamil & A. J. Moe (Eds.), *Reading research: Studies and applications*. (Twenty-Eighth Yearbook of the National Reading Conference). Clemson, S.C.: National Reading Conference, 1979.

GOUGH, P. B., & HILLINGER, M. L. Learning to read: An unnatural act. *Paper presented at the National Conference* of the Orton Society, Indianapolis, Indiana, November, 1979.

HAKES, D. T. The development of metalinguistic abilities: What develops? In S. A. Kuczaj II (Ed.), *Language acquisition: Language, cognition, and culture*. Hillsdale, N.J.: Lawrence Erlbaum Associates, in press.

HAKES, D. T., EVANS, J. S., & TUNMER, W. E. *The development of metalinguistic abilities in children*. Berlin: Springer-Verlag, 1980.

HALL, N. A. Children's awareness of segmentation in speech and print. *Reading*, 1976, *10*, 11-19.

HALLIDAY, M., & HASAN, R. *Cohesion in English*. London: Longman, 1976.

HAMMILL, D., GOODMAN, L., & WIEDERHOLD, J. L. Visual-motor processes: Can we train them? *The Reading Teacher*, 1974, *27*, 469-480.

HARGREAVES, D. J., MOLLOY, C. G., & PRATT, A. R. Social factors in conservation. *British Journal of Psychology*, 1982, *73*, 231-234.

HARRIS, P. L., KRUITHOF, A., TERWOGT, M. M., & VISSER, T. Children's detection and awareness of textual anomaly. *Journal of Experimental Child Psychology*, 1981, *31*, 212-230.

HAUGEN, E. *Bilingualism in the Americas: A bibliography and research guide*. American Dialect Society, Publication No.26. Alabama: University of Alabama Press, 1956.

HEESCHEN, V. The metalinguistic vocabulary of a speech community in the highlands of Irian Jaya (West New Guinea). In A. Sinclair, R.J. Jarvella & W. M. J. Levelt (Eds.), *The child's conception of language*. Berlin: Springer-Verlag, 1978.

HELFGOTT, J. A. Phonemic segmentation and blending skills of kindergarten children: Implications for beginning reading acquisition. *Contemporary Educational Psychology*, 1976, *1*, 157-169.

HESS, R. D., & SHIPMAN, V. C. Early experience and the socialisation of cognitive modes in children. In A. Cashdan & E. Grugeon (Eds.), *Language in education Open University course book*. London and Boston: Routledge & Kegan Paul, 1972.

HIRSH-PASEK, K., GLEITMAN, L. R., & GLEITMAN, H. What did the brain say to the mind? A study of the detection and report of ambiguity by young children. In A. Sinclair, R. J. Jarvella & W.J.M. Levelt (Eds.), *The child's conception of language*. Berlin: Springer-Verlag, 1978.

HOCKETT, C. The problem of universals in language. In J. H. Greenberg (Ed.), *Universals of language*. Cambridge, Mass.: M.I.T. Press, 1966.

HOGABOAM, T., & PERFETTI, C. Reading skill and the role of verbal experience in decoding. *Journal of Educational Psychology*, 1978, *70*, 717-729.

HOLDEN, M. H., & MacGINITIE, W. H. Children's conceptions of word boundaries in speech and print. *Journal of Educational Psychology*, 1972, *63*, 551-557.

HOOD, J., & KENDALL, J. R. A qualitative analysis of oral reading errors of reflective and impulsive second graders: A follow-up study. *Journal of Reading Behaviour*, 1975, *7*, 271-281.

HORNBY, P. A. Bilingualism: An introduction and overview. In P. A. Hornby (Ed.), *Bilingualism: Psychological social and educational implications*. New York: Academic Press, 1977.

HOWE, H. E., & HILLMAN, D. The acquisition of semantic restrictions in children. *Journal of Verbal Learning and Verbal Behavior*, 1973, *12*, 132-139.

HUGHES, M., & DONALDSON, M. The use of hiding games for studying the coordination of viewpoints. *Educational Review*, 1979, *31*, 133-140.

HUGHES, M., & GRIEVE, R. On asking children bizarre questions. *First Language*, 1980, *1*, 149-160.

HUTTENLOCHER, J. Children's language: Word-phrase relationship. *Science*, 1964, *143*, 264-265.

IANCO-WORRALL, A. D. Bilingualism and cognitive development. *Child Development*, 1972, *43*, 1390-1400.

INHELDER, B., & PIAGET, J. *The early growth of logic in the child.* London: Routledge & Kegan Paul, 1964.

ISTOMINA, Z. M. The development of voluntary memory in preschool-age children. *Soviet Psychology*, 1975, *13*, 5-64.

JAMES, S. L., & MILLER, J. F. Children's awareness of semantic constraints in sentences. *Child Development*, 1973, *44*, 69-76.

JOHNS, J. L. Metalinguistic awareness: Its growth and relationship to reading achievement. Paper presented at the International Reading Research Seminar on Linguistic Awareness and Learning to Read, Victoria, B.C., Canada, June, 1979.

JOHNS, J. L. First graders' concepts about print. *Reading Research Quarterly*, 1980, *15*, 529-549.

JUEL, C. Comparison of word identification strategies with varying context, word type, and reader skill. *Reading Research Quarterly*, 1980, *15*, 358-376.

KAGAN, J. Reflection-impulsivity and reading ability in primary grade children. *Child Development*, 1965, *36*, 609-628.

KAGAN, S., & ZAHN, G. L. Field dependence and the school achievement gap between Anglo-American children. *Journal of Educational Psychology*, 1975, *67*, 643-650.

KARMILOFF-SMITH, A. Language development after five. In P. Fletcher & M. Garman (Eds.), *Language acquisition.* Cambridge: Cambridge University Press, 1979. (a)

KARMILOFF-SMITH, A. Micro- and macrodevelopmental changes in language acquisition and other representation systems. *Cognitive Science*, 1979, *3*, 91-118. (b)

KARPOVA, S. N. The preschooler's realization of the lexical structure of speech. In F. Smith & G. A. Miller (Eds.), *The genesis of language: A psycholinguistic approach.* Cambridge, Mass.: M.I.T. Press, 1966.

KARPOVA, S. N. *The realization of the verbal composition of speech by preschool children.* Mouton: The Hague, 1977.

KÄSERMANN, M. L. *Spracherwerb und Interaktion.* Bern: Hans Huber, 1980.

KAUFMAN, A., & KAUFMAN, N. Tests built from Piaget's and Gesell's tasks as predictors of first grade achievement. *Child Development*, 1972, *43*, 521-535.

KENNEDY, D., & WEENER, P. Visual and auditory training with the cloze procedure to improve reading and listening comprehension. *Reading Research Quarterly*, 1973, *8*, 524-541.

KESSEL, F. S. The role of syntax in children's comprehension from ages six to twelve. *Monographs of the Society for Research in Child Development*, 1970, *35* (6, Serial No. 139).

KINGSTON, A. J., WEAVER, W. W., & FIGA, L. E. Experiments in children's perception of words and word boundaries. In F. P. Greene (Ed.), *Investigations relating to mature reading.* Milwaukee, Wisconsin: National Reading Conference, 1972.

KINTSCH, W., & VAN DIJK, T. A. Toward a model of text comprehension and production. *Psychological Review*, 1978, *85*, 363-394.

KIRK, S. A., McCARTHY, J. J., & KIRK, W. D. *Illinois test of psycholinguistic abilities*. Urbana, Illinois: University of Illinois Press, 1968.

KIRSNER, K., BROWN, H. L., ABROL, S., CHADHA, N. K., & SHARMA, N. K. Bilingualism and lexical representation. *Quarterly Journal of Experimental Psychology*, 1980, *32*, 585-594.

KOCHMAN, T. Towards an ethnography of black American speech behaviour. In N. E. Whitten & J. Szwed (Eds.), *Afro-American anthropology*. New York: Free Press, 1970.

KRAMSKY, J. *The word as a linguistic unit*. The Hague: Mouton, 1969.

KRAUSS, R. M., & GLUCKSBERG, S. The development of communication: Competence as a function of age. *Child Development*, 1969, *40*, 255-266.

KRAUSS, R. M., & GLUCKSBURG, S. The development of competence as a communicator. In R. A. Hoppe, G. A. Milton & E. C. Simmel (Eds.), *Early experience and the processes of socialisation*. New York: Academic Press, 1970.

KREUTZER, M. A., LEONARD, C. & FLAVELL, J. H. An interview study of children's knowledge about memory. *Monographs of the Society for Research in Child Development*, 1975, Serial No.159, *40(1)*.

LaBERGE, D., & SAMUELS, S. L. Toward a theory of automatic information processing in reading. *Cognitive Psychology*, 1974, *6*, 293-323.

LABOV, W. The logic of nonstandard English. In P. P. Giglioli (Ed.), *Language and social context*. Harmondsworth: Penguin, 1972.

LAMBERT, W. E., & TUCKER, G. R. *Bilingual education of children: The St. Lambert experiment*. Rowley, Mass.: Newbury House, 1972.

LEE, L. E. A study to determine whether bilingual pupils of high school grade are handicapped in their study of history because of vocabulary difficulties. Unpublished M.Ed. thesis; Rutgers University, 1932.

LENNEBERG, E. H. Understanding language without ability to speak: A case study. *Journal of Abnormal and Social Psychology*, 1962, *65*, 419-425.

LENNEBERG, E. H. *Biological foundations of language*. New York: Wiley, 1967.

LENNEBERG, E. H. A biological perspective of language. In R. C. Oldfield & J. C. Marshall (Eds.), *Language: Selected readings*. Harmondsworth: Penguin, 1968.

LEOPOLD, W. F. *Speech development of a bilingual child*. Evanston, Illinois: Northwestern University Press, 1949 .

LEOPOLD, W. F. Patterning in children's language learning. In S. Saporta (Ed.), *Psycholinguistics*. New York: Holt, Rinehart & Winston, 1961.

LEVELT, W.J.M., SINCLAIR, A., & JARVELLA, R. J. Causes and functions of linguistic awareness in language acquisition: Some introductory remarks. In A. Sinclair, R. J. Jarvella & W.J.M. Levelt (Eds.), *The child's conception of language*. Berlin: Springer-Verlag, 1978.

LIBERMAN, A. M. The grammars of language and speech. *Cognitive Psychology*, 1970, *1*, 301-323.

LIBERMAN, A. M., COOPER, F. S., SHANKWEILER, D. P., & STUDDERT-KENNEDY, M. Perception of the speech code. *Psychological Review*, 1967, *75*, 431-461.

LIBERMAN, I. Y. Segmentation of the spoken word and reading acquisition. *Bulletin of the Orton Society*, 1973, *23*, 65-77.

LIBERMAN, I. Y., SHANKWEILER, D., FISCHER, W. F., & CARTER, B. Explicit syllable and phoneme segmentation in the young child. *Journal of Experimental Child Psychology*, 1974, *18*, 201-212.

LIBERMAN, I. Y., SHANKWEILER, D., LIBERMAN, A. M., FOWLER, C., & FISHER, F. W. Phonetic segmentation and recoding in the beginning reader. In A. S. Reber & D. L. Scarborough (Eds.), *Toward a psychology of reading*. New York: Wiley, 1977.

LIEDKE, W. W., & NELSON, L. D. Concept formation and bilingualism. *Alberta Journal of Educational Research*, 1968, *14*, 225-232.

LIGHT, P., BUCKINGHAM, N., & ROBBINS, A. H. The conservation task as an interactional setting. *British Journal of Educational Psychology*, 1979, *49*, 304-310.

LITOWITZ, B. Learning how to make definitions. *Journal of Child Language*, 1977, *4*, 289-304.

LUNDBERG, I. Aspects of linguistic awareness related to reading. In A. Sinclair, R. J. Jarvella & W.J.M. Levelt (Eds.), *The child's conception of language*. Berlin: Springer-Verlag, 1978.

LUNZER, F., DOLAN, T., & WILKINSON, J. The effectiveness of measures of operativity, language and short-term memory in the prediction of reading and mathematical understanding. *British Journal of Educational Psychology*, 1976, *46*, 295-305.

LYONS, J. *An introduction to theoretical linguistics*. Cambridge: Cambridge University Press, 1968.

LYONS, J. Human language. In R. A. Hinde (Ed.), *Non-verbal communication*. Cambridge: Cambridge University Press, 1972.

MACE-MATLUCK, B. J. Order of acquisition: Same or different in first- and second-language learning? *The Reading Teacher*, 1979, *32*, 696-703.

MACKEY, W. F. A typology of bilingual education. *Foreign Language Annals*, 1972, *3*, 596-608.

MACKEY, W. F. The evaluation of bilingual education. In B. Spolksy & R. Cooper (Eds.), *Frontiers of bilingual education*. Rowley, Mass.: Newbury House Publishers, 1977.

MacNEILAGE, P. F., ROOTES, T. P., & CHASE, R. Speech production and perception in a patient with severe impairment of somesthetic perception and motor control. *Journal of Speech and Hearing Research*, 1967, *10*, 449-467.

MARATSOS, M. P. Children who get worse at understanding the passive: A replication of Bever. *Journal of Psycholinguistic Research*, 1974, *3*, 65-74.

MARKMAN, E. M. Children's difficulty with word-referent differentiation. *Child Development*, 1976, *47*, 742-749.

MARKMAN, E. M. Realizing that you don't understand: A preliminary investigation. *Child Development*, 1977, *48*, 986-992.

MARKMAN, E. M. Realizing that you don't understand: Elementary school children's awareness of inconsistencies. *Child Development*, 1979, *50*, 643-655.

MARKMAN, E. M. Comprehension monitoring. In W. P. Dickson (Ed.), *Children's oral communication skills*. New York: Academic Press, 1981.

MARSH, G., & MINEO, R. J. Training preschool children to recognize phonemes in words. *Journal of Educational Psychology*, 1977, *69*, 748-753.

MARSHALL, J. C., & MORTON, J. On the mechanics of EMMA. In A. Sinclair, R. J. Jarvella & W.J.M. Levelt (Eds.), *The child's conception of language*. Berlin: Springer-Verlag, 1978.

MATTINGLY, I. G. Reading, the linguistic process, and linguistic awareness. In J. F. Kavanagh & I. G. MATTINGLY (Eds.), *Language by ear and by eye*. Cambridge, Mass.: M.I.T. Press, 1972.

McGARRIGLE, J., & DONALDSON, M. Conservation accidents. *Cognition*, 1974, *3*, 341-50.

McGARRIGLE, J., GRIEVE, R., & HUGHES, M. Interpreting inclusion: A contribution to the study of the child's cognitive and linguistic development. *Journal of Experimental Child Psychology*, 1978, *26*, 528-550.

McNEILL, D. The creation of language. In R. C. Oldfield and J. C. Marshall (Eds.), *Language· Selected readings*. Harmondsworth: Penguin, 1966.

McNEIL, J. D., & STONE, J. Note on teaching children to hear separate sounds in spoken words. *Journal of Educational Psychology*, 1965, *56*, 13-15.

McNINCH, G. H. Auditory perceptual factors and measured first-grade reading achievement. *Reading Research Quarterly*, 1971, *6*, 472-492.

McNINCH, G. Awareness of aural and visual word boundary within a sample of first graders. *Perceptual and Motor Skills*, 1974, *38*, 1127-1134.

MENIG-PETERSON, C. L. The modification of the communicative behaviour in preschool children as a function of the listener's perspective. *Child Development*, 1975, *45*, 1015-1018.

MENYUK, P. *Sentences children use*. Cambridge, Mass.: M.I.T. Press, 1969.

MESSER, S. Reflection-impulsivity: A review. *Psychological Bulletin*, 1976, *83*, 1026-1052.

MILLER, S. A. Non-verbal assessment of Piagetian concepts. *Psychological Bulletin*, 1976, *83*, 405-430.

MILLER, S. A. On the generalizability of conservation: A comparison of different kinds of transformation. *British Journal of Psychology*, 1982, *73*, 221-230.

MONROE, M. *Language and how to use it: Beginning levels*. Glenview, Ill.: Scott, Foresman, 1970.

MONTESSORI, M. *The advanced Montessori method Spontaneous activity in education*, 1917. (F. Simmonds trans.) Cambridge, Mass.: Robert Bentley Inc., 1964.

MORSE, P. A. The discrimination of speech and non-speech stimuli in early infancy. *Journal of Experimental Child Psychology*, 1972, *14*, 447-492.

NELSON, K. Some evidence for the cognitive primacy of categorization and its functional basis. *Merrill-Palmer Quarterly*, 1973, *19*, 21-39.

NEILSON, I., DOCKRELL, J., & McKECHNIE, J. Does repetition of the question influence children's performance in conservation tasks? *British Journal of Developmental Psychology*, 1983, *1*.

NESDALE, A. R., PRATT, C., & TUNMER, W. E. The influence of set and story length on children's judgements of stories. Unpublished manuscript, University of Western Australia, 1982.

NEWCOMER, P., & HAMMILL, D. ITPA and academic achievement: A survey. *The Reading Teacher*, 1975, *28*, 731-741.

NISBETT, R. E., & DECAMP-WILSON, T. Telling more than we know: Verbal reports on mental processes. *Psychological Review*, 1977, *84*, 231-259.

OLSON, D. R. From utterance to text: The bias of language in speech and writing. *Harvard Educational Review*, 1977, *47*, 257-281.

ORNE, M. T. On the social psychology of the psychological experiment: With particular reference to demand characteristics and their implications. *American Psychologist*, 1962, *17*, 776-783.

OSHERSON, D., & MARKMAN, E. Language and the ability to evaluate contradictions and tautologies. *Cognition*, 1975, *3*, 213-226.

PALERMO, D. S., & MOLFESE, D. L. Language acquisition from age five onwards. *Psychological Bulletin*, 1972, *78*, 409-428.

PAPANDROPOULOU, I., & SINCLAIR, H. What is a word? Experimental study of children's ideas on grammar. *Human Development*, 1974, *17*, 241-258.

PARADIS, E. E. The appropriateness of visual discrimination exercises in reading readiness materials. *Journal of Educational Research*. 1974, *67*, 276-278.

PATTERSON, C. J., COSGROVE, J. M., & O'BRIEN, R. G. Non-verbal indicants of comprehension and non-comprehension in children. *Develomental Psychology*, 1980, *16*, 38-48.

PATTERSON, C. J., & KISTER, M. C. The development of listener skills for referential communication. In W. P. Dickson (Ed.), *Children's oral communication skills*. New York: Academic Press, 1981.

PATTERSON, C. J., MASSAD, C. M., & COSGROVE, J. M. Children's referential communication: Components of plans for effective listening. *Developmental Psychology*, 1978, *14*, 401-406.

PEAL, E., & LAMBERT, W. E. The relation of bilingualism to intelligence. *Psychological Monographs· General and Applied*, 1962, *76*, Whole No.546.

PERFETTI, C. Language comprehension and fast decoding: Some psycholinguistic prerequisites for skilled reading. In J. Guthrie (Ed.), *Cognition, curriculum, and comprehension*. Newark, Del.: International Reading Association, 1977.

PERFETTI, C., & HOGABOAM, T. The relationship between single word decoding and reading comprehension skill. *Journal of Educational Psychology*, 1975, *67*, 461-469.

PETERS, R. S. *Brett's history of psychology*. Cambridge, Massachusetts: M.I.T. Press, 1962.

PIAGET, J. *The language and thought of the child*. London: Routledge & Kegan Paul, 1926.

PIAGET, J. *Judgement and reasoning in the child*. New York: Harcourt Brace, 1928.

PIAGET, J. *The child's conception of the world*. London: Routledge & Kegan Paul, 1929.

PIAGET, J. *The child's conception of number*. New York: Humanities Press, 1952.

PIAGET, J. *The language and thought of the child* (3rd ed.). London: Routledge & Kegan Paul, 1959.

PIAGET, J. *Six psychological studies*. New York: Random House, 1967.

PIAGET, J. Piaget's theory. In P. H. Mussen (Ed.), *Carmichael's manual of child psychology*. New York: Wiley, 1970.

PIAGET, J. *La prise de conscience*. Paris: Presses Universitaires de France, 1974(a)

PIAGET, J. *Reussir et comprendre*. Paris: Presses Universitaires de France, 1974(b)

PIAGET, J. The affective unconscious and the cognitive unconscious. In B. Inhelder & H. H. Chipman (Eds.), *Piaget and his school*. New York: Springer-Verlag, 1976.

PIAGET, J. *The grasp of consciousness*. London: Routledge & Kegan Paul, 1976.

PIAGET, J., & INHELDER, B. *The child's conception of space*. London: Routledge & Kegan Paul, 1956.

PIAGET, J., INHELDER, B., & SZEMINSKA, A. *The child's conception of geometry*. London: Routledge & Kegan Paul, 1960.

POSNER, M. I., & SNYDER, C. R. R. Attention and cognitive control. In R. L. Solso (Ed.), *Information processing and cognition*. Hillsdale, N.J.: Lawrence Erlbaum Associates, 1975.

PRATT, C. The awareness and communication of non-communication. In T. Le & M. McCausland (Eds.), *Child language development: Theory into practice*. Launceston, Tasmania, 1980.

PRATT, C. Met a problem with metacognition. In A. R. Nesdale, C. Pratt, R. Grieve, J. Field, D. Illingworth & J. Hogben (Eds.), *Advances in child development: Theory and research*, Perth, W.A.: N.C.C.D., 1981.

PRATT, C. Young children's judgement and production of message adequacy. Unpublished manuscript, University of Western Australia, 1982.

PRATT, C., TUNMER, W. E., & BOWEY, J. Children's capacity to correct grammatical violations in sentences. *Journal of Child Language*, in press.

PRATT, C. TUNMER, W. E., & NESDALE, A. R. The effects of story type on children's judgements of stories. Unpublished manuscript, University of Western Australia, 1982.

PRATT, M. W., & BATES, K. R. Young editors: Preschoolers' evaluation and production of ambiguous messages. *Developmental Psychology*, 1982, *18*, 30-42.

RAMIREZ, M. Cognitive styles and cultural democracy in education. *Social Science Quarterly*, 1973, *53*, 895-904.

RAMIREZ, M., & CASTANEDA, A. *Cultural democracy, bicognitive development, and education*. New York: Academic Press, 1974.

RAMIREZ, M., CASTANEDA, A., & HEROLD, P. L. The relationship of acculturation to cognitive style among Mexican Americans. *Journal of Cross-Cultural Psychology*, 1974, *5*, 425-433.

REID, J. F. Learning to think about reading. *Educational Research*, 1966, *9*, 56-62.

RESNICK, L. B. Relations between perceptual and syntactic control in oral reading. *Journal of Educational Psychology*, 1970, *61*, 382-385.

ROBINSON, E. J., & ROBINSON, W. P. The young child's understanding of communication. *Developmental Psychology*, 1976, *12*, 328-333. (a)

ROBINSON, E. J., & ROBINSON, W. P. Developmental changes in the child's explanation of communication failure. *Australian Journal of Psychology*, 1976, *28*, 155-165. (b)

ROBINSON, E. J., & ROBINSON, W. P. Children's explanations of communication failure and the inadequacy of the misunderstood message. *Developmental Psychology*, 1977, *13*, 156-161. (a)

ROBINSON, E. J., & ROBINSON, W. P. Development in the understanding of causes of success and failure in verbal communication. *Cognition*, 1977, *5*, 363-378. (b)

ROBINSON, E. J., & ROBINSON, W. P. Development of understanding about communication: Message inadequacy and its role in causing communication failure. *Genetic Psychological Monographs*, 1978, *98*, 233-279. (a)

ROBINSON, E. J., & ROBINSON, W. P. The relationship between children's explanations of communication failure and their ability deliberately to give bad messages. *British Journal of Social and Clinical Psychology*, 1978, *17*, 219-225. (b)

ROBINSON, E. J., & ROBINSON, W. P. Egocentrism in verbal referential communication. In M. V. Cox (Ed.), *Are young children egocentric?* London: Batsford Academic and Educational Ltd., 1980.

ROBINSON, E. J., & ROBINSON, W. P. Ways of reacting to communication failure in relation to the development of the child's understanding about verbal communication. *European Journal of Social Psychology*, 1981, *11*, 189-208.

ROBINSON, E. J., & ROBINSON, W. P. Children's uncertainty about the interpretation of ambiguous messages. *Journal of Experimental Child Psychology*, in press. (a)

ROBINSON, E. J., & ROBINSON, W. P. The advancement of children's verbal referential communication skills: The role of metacognitive guidance. *International Journal of Behavioural Development*, in press. (b)

ROBINSON, P. *Language management in education: The Australian context.* Sydney: Allen & Unwin, 1978.

ROBINSON, W. P. *Language and social behaviour.* Harmondsworth: Penguin, 1972.

ROLL, S. Reversibility training and stimulus desirability as factors in conservation of number. *Child Development*, 1970, *41*, 501-507.

ROSE, S. A., & BLANK, M. The potency of context in children's cognition: An illustration through conservation. *Child Development*, 1974, *45*, 499-502.

ROSEN, C. An experimental study of visual perceptual training and reading achievement in first grade. *Perceptual and Motor Skills*, 1966, *22*, 979-986.

ROSNER, J. *Phonic analysis training and beginning reading skills* (Publication series No.19). Pittsburgh, Pa.: University of Pittsburgh, Learning Research and Development Center, 1971.

ROSNER, J. Auditory analysis training with prereaders. *The Reading Teacher*, 1974, *27*, 378-384.

ROSNER, J., & SIMON, D. P. The auditory analysis test: An initial report. *Journal of Learning Disabilities*, 1971, *4*, 384-392.

ROUSSEAU, J. J. *Emile*, 1762. (B. Foxley trans.) London: Dent & Sons, 1948.

ROZIN, P., BRESSMAN, B., & TAFT, M. Do children understand the basic relationship between speech and writing? The mow-motor cycle test. *Journal of Reading Behaviour*, 1974, *6*, 327-334.

ROZIN, P., & GLEITMAN, L. R. The structure and acquisition of reading II: The reading process and the acquisition of the alphabetic principle. In A. S. Reber & D. L. Scarborough (Eds.), *Toward a psychology of reading*. Hillsdale, N. J.: Lawrence Erlbaum Associates, 1977.

RUDDELL, R. B. The effect of similarity of oral and written patterns of language structure on reading comprehension. *Elementary English*, 1965, *43*, 403-410.

SACHS, J., & DEVIN, J. Young children's use of age-appropriate speech styles in social interaction and role-playing. *Journal of Child Language*, 1976, *3*, 81-98.

SAMUELS, S. J. Modes of word recognition. In H. Singer & R. Ruddell (Eds.), *Theoretical models and processes of reading*. Newark, Del.: International Reading Association, 1970.

SAMUELS, S. J. Letter-name versus letter-sound knowledge in learning to read. *The Reading Teacher*, 1971, *24*, 604-608.

SAMUELS, S. J. The effect of letter-name knowledge on learning to read. *American Educational Research Journal*, 1972, *9*, 65-74.

SAMUELS, S. J., & JEFFREY, W. D. Discriminability of words and letter cues used in learning to read. *Journal of Educational Psychology*, 1966, *57*, 337-340.

SAMUELS, S. J., & TURNURE, J. E. Attention and reading achievement in first grade boys and girls. *Journal of Educational Psychology*, 1974, *66*, 29-32.

SAPIR, E. *Language*. New York: Harcourt Brace & World, 1921.

SAVIC, S. *How twins learn to talk*. New York: Academic Press, 1980.

SAVIN, H. B. What the child knows about speech when he starts to learn to read. In J. F. Kavanagh & I. G. Mattingly (Eds.), *Language by ear and by eye*. Cambridge, Mass.: M.I.T. Press, 1972.

SCHILLER, A., MONROE, M., NICHOLS, R., JENKINS, W., & HUCK, C. *Language and how to use it. Book 1* (Teacher's edition). Glenview, Ill.: Scott, Foresman, 1973.

SCHNEIDER, W., & SHIFFRIN, R. M. Controlled and automatic human information processing: I. Detection, search and attention. *Psychological Review*, 1977, *84*, 1-66.

SCHOLL, D. M., & RYAN, E. B. Child judgments of sentences varying in grammatical complexity. *Journal of Experimental Child Psychology*, 1975, *20*, 274-285.

SCHOLL, D. M., & RYAN, E. B. Development of metalinguistic performance in the early school years. *Language and Speech*, 1980, *23*, 199-211.

SCOLLON, R. *Conversation with a one year old: A case study of the developmental foundation of language*. Honolulu: University Press of Hawaii, 1976.

SHANKWEILER, D., & LIBERMAN, I. Y. Misreading: A search for causes. In J. F. Kavanagh & I. G. Mattingly (Eds.), *Language by ear and by eye*. Cambridge, Mass.: M.I.T. Press, 1972.

SHATZ, M. Preschoolers' ability to take account of others in a toy selection task. Unpublished masters thesis, University of Pennsylvania, 1973.

SHATZ, M., & GELMAN, R. The development of communication skills: Modification in the speech of young children as a function of listener.

Monographs of the Society for Research in Child Development, 1973, *38*, (5, Serial No. 152), 1-38.

SHIFFRIN, R. M., & SCHNEIDER, W. Controlled and automatic human information processing: II. Perceptual learning, automatic attending, and a general theory. *Psychological Review*, 1977, *84*, 127-190.

SHULTZ, T. R. Development of the appreciation of riddles. *Child Development*, 1974, *45*, 100-105.

SHULTZ, T. R., & PILON, R. Development of the ability to detect linguistic ambiguity. *Child Development*, 1973, *44*, 728-733.

SIDOWSKI, J. B., & LOCKARD, R. B. Some preliminary considerations in research. In J. B. Sidowski (Ed.), *Experimental methods and instrumentation in psychology*. New York: McGraw-Hill, 1968.

SILBERBERG, N. E., SILBERBERG, M. D., & IVERSON, I. A. The effects of kindergarten instruction in alphabet and numbers on first-grade reading. *Journal of Learning Disabilities*, 1972, *5*, 7-12.

SINCLAIR, H. Conceptualization and awareness in Piaget's theory and its relevance to the child's conception of language. In A. Sinclair, R. J. Jarvella & W.J.M. Levelt (Eds.), *The child's conception of language*. Berlin: Springer-Verlag, 1978.

SLOBIN, D. I. A case study of early language awareness. In A. Sinclair, R.J. Jarvella & W.J.M. Levelt (Eds.), *The child's conception of language*. Berlin: Springer-Verlag, 1978.

SMITH, E. R., & MILLER, F. D. Limits on perception of cognitive processes: A reply to Nisbett and Wilson. *Psychological Review*, 1978, *85*, 355-362.

SMITH, F. *Understanding reading: A psycholinguistic analysis of reading and learning to read*. New York: Holt Rinehart & Winston, 1971.

SNYDER, A. D. Notes on the talk of a two-and-a-half year old boy. *Pedagogical Seminary*, 1914, *21*, 412-424.

STANOVICH, K. E. Toward an interactive-compensatory model of individual differences in the development of reading fluency. *Reading Research Quarterly*, 1980, *16*, 32-71.

STANOVICH, K. E., WEST, R. F., & FEEMAN, D. J. A longitudinal study of sentence context effects in second-grade children: Tests of an interactive-compensatory model. *Journal of Experimental Child Psychology*, 1981, *32*, 185-199.

STEBBINS, L. B., ST. PIERRE, R. G., PROPER, E. C., ANDERSON, R. B., CERVA, T. R., & KENNEDY, M. M. *Evaluation as experimentation: A planned variation model*. Vol.4-A. Cambridge, Mass.: Abt Associates, 1977.

STUART, I. R. Perceptual style and reading ability: Implications for an instructional approach. *Perceptual and Motor Skills*, 1967, *24*, 135-138.

SUTCLIFFE, J. P. On the role of "instructions to the subject" in psychological experiments. *American Psychologist*, 1972, *27*, 755-758.

SUTTON-SMITH, B. A developmental structural account of riddles. In B. Kirshenblatt-Gimblett (Ed.), *Speech play*. Philadelphia: University of Pennsylvania Press, 1976, pp.111-119.

SWAIN, M. French immersion: Early, late or partial? In S. T. Carey (Ed.), *The Canadian Modern Language Review*, 1978, *34*, 557-585.

TORRANCE, E. P., GOWAN, J. C., WU, J. M., & ALIOTTI, N. C. Creative functioning of monolingual and bilingual children in Singapore. *Journal of Educational Psychology*, 1970, *61*, 71-75.

TUCKER, G. R. The linguistic perspective. In *Bilingual education: Current perspectives, Vol.2. Linguistics*. Arlington, Va.: Center for Applied Linguistics, 1977.

TUNMER, W. E., & BOWEY, J. A. The development of word segmentation skills in children. In A. R. Nesdale, C. Pratt, R. Grieve, J. Field, D. Illingworth and J. Hogben (Eds.), *Advances in child development: Theory and research*. Perth, W.A.: N.C.C.D., 1981.

TUNMER, W. E., BOWEY, J.A., & GRIEVE, R. The development of young children's awareness of the word as a unit of spoken language. *Journal of Psycholinguistic Research*, in press.

TUNMER, W. E., & FLETCHER, C. M. The relationship between conceptual tempo, phonological awareness, and word recognition in beginning readers. *Journal of Reading Behavior*, 1981, *13*, 173-186.

TUNMER, W. E., & NESDALE, A. R. The effects of digraphs and pseudowords on phonemic segmentation in young children. *Journal of Applied Psycholinguistics*, 1982, *3*, 299-311.

TUNMER, W. E., NESDALE, A. R., & PRATT, C. The development of young children's awareness of logical inconsistencies. *Journal of Experimental Child Psychology*, in press.

U.S. COMMISSION ON CIVIL RIGHTS. *The unfinished education*. Washington, D.C.: U.S. Government Printing Office, 1971.

VALTIN, R. Increasing awareness of linguistic awareness in research of beginning reading and dyslexia. Paper presented at the International Reading Research Seminar on Linguistic Awareness and Learning to Read, Victoria, B.C., Canada, June, 1979.

VENEZKY, R. L. *The structure of English orthography*. Paris: Mouton, 1970.

VENEZKY, R. L. *Theoretical and experimental bases for teaching reading*. The Hague: Mouton, 1976.

VENEZKY, R. L. Reading acquisition: The occult and the obscure. In F.B. Murray & J.J. Pilulski (Eds.), *The acquisition of reading: Cognitive, linguistic and perceptual prerequisites*. Baltimore: University Park Press, 1978.

VYGOTSKY, L. S. *Thought and language*. Cambridge, Mass.: M.I.T. Press, 1962.

WALES, R. J. Children's sentences make sense of the world. In F. Bresson (Ed.), *Problems actuels en psycholinguistique*. Paris: P.U.F., 1974.

WALLACH, L., & SPROTT, R. L. Inducing number conservation in childen. *Child Development*, 1964, *35*, 1057-1071.

WALLACH, L., WALL, A. J., & ANDERSON,L. Number conservation: The roles of reversibility, addition-subtraction and misleading perceptual cues. *Child Development*, 1967, *38*, 425-442.

WALLACH, L., WALLACH, M. A., DOZIER, M. G., & KAPLAN, N. W. Poor children learning to read do not have trouble with auditory discrimination but do have trouble with phoneme recognition. *Journal of Educational Psychology*, 1977, *69*, 36-39.

WALLACH, M. A., & WALLACH, L. *Teaching all children to read*. Chicago: University of Chicago Press, 1976.

WEAVER, P. Improving reading comprehension: Effects of sentence organization instruction. *Reading Research Quarterly*, 1979, *15*, 127-146.

WEAVER, P., & SHONKOFF, P. *Research within research: A research-guided response to concerns of reading educators*. St. Louis: CEMREL, 1978.

WEBER, R. M. First grader's use of grammatical context in reading. In H. Levin & J. Williams (Eds.), *Basic studies in reading.* New York: Basic Books, 1970.

WEINRICH, V. *Languages in contact.* The Hague: Mouton, 1953.

WEINSTEIN, R., & RABINOVICH, M. S. Sentence structure and retention in good and poor readers. *Journal of Educational Psychology*, 1971, *62*, 25-30.

WEIR, R. H. *Language in the crib.* The Hague: Mouton, 1962.

WEIR, R. H. Some questions on the child's learning of phonology. In F. Smith & G. A. Miller (Eds.), *The genesis of language.* Cambridge, Mass.: M.I.T. Press, 1966.

WELLS, C. G. *Learning through interaction: The study of language development.* Cambridge: Cambridge University Press, 1981.

WILLIAMS, J. Teaching decoding with an emphasis on phoneme analysis and phoneme blending. *Journal of Educational Psychology*, 1980, *72*, 1-15.

WHITEHURST, G. J. Commentary on J. H. Flavell, J. R. Speer, F. L. Green & D. L. August. The development of comprehension monitoring and knowledge about communication. *Monographs of the Society for Research in Child Development*, 1981, Serial No.192, *46(5)*, pp.59-65.

WIENER, M., DEVOE, S., RUBINOW, S., & GELLER, J. Non-verbal behaviour and non-verbal communication. *Psychological Review*, 1972, *79*, 185-214.

WINGERT, R. C. Evaluation of a readiness training program. *The Reading Teacher*, 1969, *22*, 325-328.

Author Index

Subject Index

Springer Series in Language and Communication

Editor: **W. J. M. Levelt**

Volume 1
W. Klein, N. Dittmar
Developing Grammars
The Acquisition of German Syntax by Foreign Workers
1979. 9 figures, 38 tables. X, 222 Seiten
ISBN 3-540-09580-2

Volume 2
The Child's Conception of Language
Editors: **A. Sinclair, R. J. Jarvella, W. J. M. Levelt**
1978. 9 figures, 5 tables. IX, 268 pages
ISBN 3-540-09153-X

Volume 3
M. Miller
The Logic of Language Development in Early Childhood
Translated from the German by R.T.King
1979. 1 figure, 30 tables. XVI, 478 pages
ISBN 3-540-09606-X

Volume 4
L. G. M. Noordman
Inferring from Language
With a Foreword by H. H. Clark
1979. 4 figures, 25 tables. XII, 170 pages
ISBN 3-540-09386-9

Volume 5
W. Noordman-Vonk
Retrieval from Semantic Memory
With a Foreword by J. C. Marshall
1979. 10 figures, 19 tables. XII, 97 pages
ISBN 3-540-09219-6

Volume 6
Semantics from Different Points of View
Editors: **R. Bäuerle, U. Egli, A. v. Stechow**
1979. 15 figures, 7 tables. VIII, 419 pages
ISBN 3-540-09676-0

Volume 7
C. E. Osgood
Lectures on Language Performance
1980. 31 figures, 33 tables. XI, 276 pages
ISBN 3-540-09901-8

Volume 8
T. Ballmer, W. Brennenstuhl
Speech Act Classification
A Study in the Lexical Analysis of English Speech Activity Verbs
1981. 4 figures. X, 274 pages
ISBN 3-540-10294-9

Volume 9
D. T. Hakes
The Development of Metalinguistic Abilities in Children
In collaboration with J. S. Evans and W. Tunmer
1980. 6 figures, 8 tables. X, 119 pages
ISBN 3-540-10295-7

Volume 10
R. Narasimhan
Modelling Language Behaviour
1981. 3 figures. XVI, 217 pages
ISBN 3-540-10513-1

Springer-Verlag
Berlin
Heidelberg
New York
Tokyo

MIX
Papier aus verantwortungsvollen Quellen
Paper from responsible sources
FSC® C105338

If you have any concerns about our products,
you can contact us on
ProductSafety@springernature.com

In case Publisher is established outside the EU,
the EU authorized representative is:
Springer Nature Customer Service Center GmbH
Europaplatz 3, 69115 Heidelberg, Germany

Printed by Libri Plureos GmbH
in Hamburg, Germany